A New Reality

Life Below the "E" State

Second Edition

C. R. Boretsky

Order this book online at www.trafford.com
or email orders@trafford.com

Most Trafford titles are also available at major online book retailers.

Cover Photos Courtesy of NASA

Upper Photo: Hurricane Fran, September 4, 1996

Image Credit: NOAA GOES-8 1715 UTC NASA
                Goddard Laboratory for Atmospheres

Lower Photo: Dusty Spiral Galaxy NGC 4414

Image Credit: The Hubble Heritage Team (AURA/STScI/NASA)

Printed in the United States of America.

ISBN: 978-1-4669-7924-6 (sc)
ISBN: 978-1-4669-7923-9 (hc)
ISBN: 978-1-4669-7922-2 (e)

Library of Congress Control Number: 2013901728

Trafford rev. 01/21/2015

 www.trafford.com

North America & international
toll-free: 1 888 232 4444 (USA & Canada)
fax: 812 355 4082

To my wife, Kimberly,

without whose patience, support, and love

I could not have completed this project

Contents

The New Preface

In the years 1995-1997, I decided to write down ideas that had been with me for many years. That's kind of the way I am. I think about a problem that needs a solution, and I have to eventually put it down on paper. After I finished this book, I set it aside and stopped thinking about fixing the world. A few years passed, and before I knew it, a lot of the things I had thought about and had written about started to affect our world. That's when I added the chapter on 9/11. I showed my book to a few friends, and they encouraged me to try to publish my thoughts. So I did. The venue I chose was Trafford Publishing, a Canadian Company that specializes in self-publishing. I sent my manuscript off in the fall of 2003. In the fall of 2004, they finally got my book in print. Yeehah! Now everyone will have a chance to read my book. Little did I realize that everyone on the planet was trying to get their book published just as I was. My poor little book was lost in the abyss. A few copies were sold, but not enough to pay for a decent meal at a fast-food chain.

It is now the winter of 2012. As you may have noticed, the font I am using is different from that of the following text. I have chosen to revisit my own book and make updates and changes, which have taken place in the past decade. I have also enhanced my own version of reality as it pertains to the formation of the universe. Revelations don't occur overnight, you know. Many of the topics you are going to read about might seem like old news now. But I assure you, they were not old news seventeen years ago when I began writing about them. I hope you will keep an open mind as you read my book.

Many of these topics will make you feel uneasy. That's okay. That's what change is all about. It's time for us to move forward into the next phase of human development. An ascension of humankind, if you will. I think that we are ready. But I also think that we are faced with tremendous hurdles. Letting go of the past is never easy. Remembering and learning from the past will be our salvation.

PREFACE
to the First Edition

Who I am is not important. What I have to say may be the most important gift that I can give to the world. This is a compilation of thoughts and beliefs that I have been developing since I was a young adult. Many of these thoughts and beliefs will be familiar to many of you since I believe that we share a common consciousness. My early studies in school concentrated mostly on the sciences. I would consider myself an average student, and I have the grades to prove it. What I found myself doing throughout most of my studies was always asking the question *why*. I must have asked often enough or had a little help along the way, but slowly I started getting answers. They weren't always the ones I wanted to hear, but they made a lot of sense years later. A lot of what I have to say involves something very personal to most of us—namely, religion. This was one area that I really asked the question *why* many times. I don't want you to think right from the beginning of this text that I am going to try to convert you to some obscure religious cult or convince you that I have all the answers. I don't. What I can tell you is this: I will shed some light on a world that we still have a lot to learn about. Hopefully, when you finish this book, you too may see the world in a little different way. You may start to see our world as I see it through my eyes. It's time for the world to have a new perspective on reality.

Chapter One

In the Beginning

"In the beginning, God created heaven and earth" has been a paradigm that has lasted through the centuries. It has been the foundation of Jewish and Christian faiths since writings of the Bible were scribed in their original text. But accepting that there is a beginning must also imply that there is an ending. An alpha and an omega, if you will. But an ending to what? To time. To life. To the earth. To all these.

Let me pose for a moment that there is no end to any of these events. Let me instead say that there is no limit to time, life, or the end of the earth. Would you even think that this is a possibility? I would like to tell you that, in my world, it is not. I view all things as transitional. There is no beginning, and there is no end. It's what's in between that is the interesting part. This is where we as human beings exist, and of course, that's where our lives come into the scheme of things.

In figure 1, I introduce life as we know it, as consisting of two basic concepts—a physical realm and a spiritual realm. Above these concepts are the biblical references of the alpha and omega or Satan and God. Notice that I have indicated that the pursuit of the physical realm tends toward degradation while the pursuit of the spiritual realm tends toward enlightenment. In the middle is us, a synergy of the realms.

As you read further into this book, I will shed additional light as to the meaning of it all.

So what is a beginning? Let's begin with the birth of a child as an appropriate starting point or beginning. Firstly, is this a beginning at all? I think most of us would say yes. I would say no. Why? Because the birth of a child is a culmination of events that took place far ahead of the birth event in the delivery room. Before there was a child, there was a mother and a father, who would produce an egg and fertilize that egg, respectively. In order for them to perform that conception, they both would have to receive the proper nutrients from the earth to enable their bodies to produce healthy eggs and sperm. The nutrients would be drawn from the soil, which, in reality, is the decomposition of organic and inorganic material, the origin of the organic material being waste products from living and decomposing plants and animals.

I know this is pretty basic biology, but my emphasis is on the fact that many events must take place prior to an event, which we would term a beginning. Consider the death of an adult as an ending. I would say that the life and death of any living organism is only a transition from the soil, back to the soil. Ashes to ashes and dust to dust, if you will. I'm sure by this time many of you are asking, "What about the soul?" I will only tell you at this time that our souls are no exception to the transitional state of existence. They also have no beginning or ending.

Let me shift the discussion on beginnings and endings to a more abstract format, that of mathematics. In mathematics, we make many assumptions or postulates, which lead us to a logical conclusion. When studying numbers and fractions of numbers, we found that for every numerical value, there was always another numerical value either larger or smaller. In fact, the numbers get so large or so small we assume they go on forever, and of course, they do. In advanced mathematics, equations take on properties known as either convergence or divergence. It simply means that the resultant value either becomes extremely large or extremely small. Mathematicians then postulate that a divergent equation will eventually result in an infinitely large solution or infinity, and a convergent equation will eventually result in an infinitely small solution or zero.

So what does this have to do with anything? Simple. Mathematics assumes infinite values or infinity as a solution to explaining the physical world. There is no beginning or ending to numerical values. The physical and spiritual worlds have no boundaries, no stops and starts, no beginnings or endings. The ones that we imagine as beginnings are only transitions from another form. The good news is that all events in the universe transition in cycles. It is the cycles in our lives that I believe give us a sense of stability.

So here is my first set of rules for the day:
1. There is no such thing as a beginning or ending.
2. All things are in transition.
3. All transitions are cyclical.

You may ask the question, "If the universe is infinite, how can we exist on a finite planet?" Mathematics once again provides the solution. A line on a two-dimensional plane is considered to have an infinite number of points along that line. Each point, or series of points, has a specific value or set of values. The points can be grouped and segmented and are unique among themselves even though they exist on an infinite line. Earth, the sun, the solar system, and the Milky Way all exist within an infinite universe. All can be defined as a point or set of points in space. All are unique. We on Earth are finite portions of matter in an infinite sea of matter.

Figure 1

Life

The Alpha
(Satan)
The Physical Realm

The Omega
(God)
The Spiritual Realm

It's what's in between that
counts! (Us)

Chapter Two

Mother Earth

Sir Isaac Newton was a brilliant man. Why? Because he formulated three fundamental rules of physics, one of which stated, "Matter can neither be created nor destroyed." I remember reading in my college freshman physics book a wonderful sentence at the beginning of one of the early chapters. It stated that all the matter on Earth has been here since its creation. Everything that exists on Earth today is made up from the same matter as earlier civilizations and living organisms. Examine the world around you today. Absolutely everything you see, touch, smell, and taste is made up of ageless atoms and molecules that have been a part of Earth since it was created from cosmic dust. And by the way, where did the cosmic dust come from originally? I'll get to that in a later chapter.

What is not particularly obvious is that the world that we are familiar with during our lifetime is merely a temporary state of matter. Since most of our surroundings will be here well beyond our life spans, it is difficult to imagine Earth in any other form.

The atoms of matter contained in our bodies were once other living and nonliving things dwelling upon this planet. When we die, our bodies will then help to make up living organisms in our future. The cycle continues.

With this knowledge come some rather profound social implications. Remember that our predecessors are now a part of our living bodies. Who were our predecessors? Those of us in North America don't need to look too far into the past to find Native Americans, French, Spanish, and African Americans as just a sample of other cultures sharing the same land before us. As difficult as it may seem to you, we now exist from some of the same matter as these people once did. We share their atoms. We are all physically related to every living thing before us. Knowing this, prejudicial behavior toward your fellow human being seems a little ridiculous.

We share the same planet.

We are the same planet.

Welcome to Mother Earth.

Chapter Three

The Perfect Computer: The Brain

You might think that it is kind of strange to think of your brain as a computer, but it truly is. We process incredible amounts of information each day, which results in a daily pattern or routine. Of course, some days don't always end up in the routine manner in which we would like. In these instances, we now have to rely on our experiences. To do that, first we have to have some.

From our birth, our mind is subjected to a wide range of inputs. Some come from our home environments, and some come from the environments of other families, outdoor activities, or formal education. Each one serves to program our brain to function in a logical and socially acceptable manner. What determines that acceptable manner depends on the source of the input. In other words, how we think depends on others around us and the physical world around us. Additionally, since our brain functions as a result of chemical reactions, we must also consider the effects of nutrition on the thought process.

Neural pathways in the brain are formed from repetition and continual use. Without proper nutrition to sustain cell growth, neural pathways in our brains will not properly process information necessary to perform complex tasks.

Where I am heading with all this discussion is the foundation for discovering abnormal behavior and brain dysfunction. In other words, why can't we all think alike and behave in socially acceptable manners? The answers are really quite obvious but not simple to remedy.

The ability for our brains to function properly depends on a "normal" home environment coupled with proper nutrition. Proper nutrition also includes the nutrition of the mother during pregnancy, and the lack of chemical contaminants in her bloodstream. Also, the mother's ability to provide nutrients to the growing child depends on the mother's mother. Every generation is dependent on the previous ones to provide a healthy foundation for proper growth.

Drugs, food additives, preservatives, pesticides, herbicides, and unbalanced diets will eventually result in inferior cell growth in children from generation to generation. This includes the development of the brain, central nervous system, bone and muscle growth, and the ability of women to give birth to healthy babies.

After decades of human consumption of these products, we are only now beginning to understand the long-term consequences of ingesting food additives. Although many additives help to prevent the ingestion of toxic bacteria, such as butylated hydroxytoluene (BHT) and butylated hydroxyanisole (BHA), many others, such as food colorings, may do more harm than good. Nitrites, which are widely used as preservatives, can lead to certain forms of cancer if converted to nitrosamines within the body or if consumed in large enough quantities. Aspartame, a common sugar substitute in soft drinks, chewing gum and desserts has now been linked to multiple health risks including brain damage, depression, vision and auditory loss and non-Hodgkins lymphoma. Those of us who consume the majority of our diets from processed food sources are multiplying the ill effects of these additives.

In recent years, we have seen the introduction of organically produced food products, which are supposed to be lower in pesticides and nitrates (fertilizer). Unfortunately, through extensive testing, these products have not been found to be appreciably lower in harmful chemicals. What makes these food products potentially more beneficial for consumption is the use of vitamin E as a food preservative in lieu of other chemical compounds. Flash freezing has also been helpful in reducing the amounts of chemical preservatives required to minimize the spoilage in produce and fish. The bottom line is, buy fresh, cook fresh, and avoid an overabundance of processed foods. And yes, this means eating more fruits and vegetables and avoiding processed sugar. Mom was right after all!

Now comes the really hard part. How do we provide a "normal" home environment? I have to use quotations around the word *normal* since there really doesn't seem to be a standard for normality. If your normal home environment includes spouse abuse, drug abuse, or child abuse, then your brain will perceive this as normal behavior. Your mind will become programmed to accept this as socially acceptable. The neural network will be established in your brain. If your home environment includes praise for good behavior, physical exercise as a daily routine, and nutritional education, your brain will develop to accept these as normal behaviors.

The priorities we establish as adults are directly related to our normal home environment and the nutritional needs of our bodies. Abnormal behavior and dysfunction now become more a result of home environment and nutrition than physical abnormalities. (Although the latter can produce further-reaching damage.)

Once we establish genetic disorders within our society, we are apt to face some frightening revelations. As much as we would like to distance ourselves from the personalities of our parents, we are destined to exhibit neurological patterns similar to our parents. Hormonal development, musculature, height, weight, and skin and hair color are all genetically transmitted to our offspring. I refer to this trait as "genetic memory."

One more trait that we often dispel, but must face, is demeanor. In animal breeding, it is known as temperament.

For centuries we have selected and bred domesticated animals based on their physical attributes and temperament. We have bypassed natural selection and have "fine-tuned" various species to our liking. The human species, for the most part, has been left to its own to procreate unfettered or freely. This does not pose a problem genetically until the world becomes saturated with human beings, as it is now.

Countries with high education standards and strong industrial bases have faced this dilemma by instituting sophisticated birth control measures. Family size is limited to coincide with mortality rates of their countries. Smaller family size also equates to lower financial burden for the parents and a more intimate interaction of the parents with the children. Fertile lands are left untouched by human encroachment and thereby produce food quantities equal to consumption.

Recent studies have shown that as education levels of females in a society have increased, birthrates have dropped. Women who stay in school longer are less likely to become pregnant than those women who drop out of school or only finish a primary education. When they do have children, they are older, more affluent, and tend to only conceive one to two children, total.

Underdeveloped countries with uneducated populations reproduce at a rate unrelated to mortality rates. Large family size is a plus, enabling young children to drop out of school, secure a job, and bring money back to the parents. Jobs are scarce since so many people need jobs. Pay is low since the workforce is large, and the workers are limited due to their poor education. Industrialization is minimal. Parental supervision is minimal. Teen pregnancies abound. Fertile lands are needed for housing, resulting in inadequate food production for an increasing population. Disease and starvation ensue.

The earth has a finite limit to food production and freshwater supplies, so it follows that the more humans that inhabit the earth, the fewer natural resources will be available for consumption. With fewer humans, a greater standard of living will prevail (food, space, and energy). With more humans, a reduced standard of living will prevail. In short, more for less, less for more.

Nature chooses its own solution to curb human overpopulation. What we in the United States are currently coming to realize is that young segments of our own society are behaving like underdeveloped nations. Lack of parental supervision results in poor education, resulting in early dropouts, resulting in low-paying jobs or no job at all, resulting in teen pregnancies, resulting eventually into an unqualified workforce incapable of sustaining an industrialized economy. Individuals caught up in this cycle seem to gravitate to others in the same dilemma. Their demeanors are similar. They tend to produce offspring with similar temperaments. Drug and alcohol abuse is rampant. Children of these parents many times are born with preexisting cocaine addictions or *fetal alcohol syndrome*. Genetic disorder has been established. Birth control is nonexistent, so family size tends to be large. Interpersonal relationships between the mother and father are severely strained due to limited scholastic or spiritual development. Divorce usually ensues, assuming wedding vows were ever exchanged. Drive and ambition tend to be low due to a general feeling of hopelessness. Those that overcome their plight may find it difficult to succeed because of limited cognitive ability. Hence, the cycle continues.

Over time, these small segments will become large enough to create an enormous strain on our resources. There will become an ever-increasing division between the educated and uneducated. Civil unrest and crime will pervade our society. How we change this pattern to influence our future will be discussed in a future chapter.

Since I wrote this chapter fifteen years ago, do you think that this prediction was accurate or not?

Chapter Four

Truth, Facts, and Perceptions

Our legal system is based on seeking the truth and providing justice to those who have been wronged or killed. The basis for determining who is at fault or if fault exists depends on the compilation of facts. From the facts will then come the perception of guilt or innocence. This may all sound fairly straightforward until you realize that fact is based on human perception.

Is human perception capable of determining fact? That depends on truth. I understand that this all sounds like a lot of tail chasing, but truth is the most important concept to humankind's moral, ethical, and social development. Truth will establish our foundation for technical advancement and religious beliefs. Truth is also the most difficult concept to master.

Science and scientific research depend upon previous generations of work and their concept of truth to make breakthroughs in tomorrow's advancements. We rely heavily on the perceptions of *Aristotle, Copernicus, Galileo, Kepler, Newton, Planck*, and *Einstein* to define our world in ways that we can comprehend. It is through their concept of reality that we define our own reality. Is our current reality really the truth? In other words, is the way in which we perceive our world today, without question, the absolute truth?

In order to answer this question, we need only look at our past perceptions of our world. Greek scholars in the third century BC would have convinced us that the world was flat. They also would have insisted that the sun revolved around Earth, that there was only one continent, and that the world was made up of five things: water, earth, fire, air, and *quintessence*. Roughly nineteen hundred years later, scholars would insist that the world was round, revolved around the sun, and was still located at the center of the universe. By the late seventeen hundreds, scholars would realize that our world is not in the center of the universe; in fact, Earth is not even centered in our own galaxy.

In each one of these examples, the reality of the time was established by the compilation of fact after fact, each one supporting the other so as to result in the essence of truth. But what motive could there be for hanging on to a reality that just didn't match up with the evidence? Why does mankind resist change? What are we afraid of? The answer is simply that mankind has an intense desire to believe that it is singularly special. We want to be unique, and are willing to do whatever it takes, including altering the evidence, to prove to ourselves that we are right.

Courts of law are a prime example of evidence tampering. By tampering, I am suggesting that it is not important what the facts of the case are but, instead, what the perceptions of the jurists are. Defense attorneys are notorious for selecting jurors who they believe can be convinced to view evidence in a way that will make it appear as though their client is innocent. It is incumbent for the defense attorney to skew facts in a case in order to vindicate a guilty client. For example, "The bloody knife with both the victim's and perpetrator's DNA could not possibly be the murder weapon because the blood samples were tainted at the forensic lab with preservatives. Therefore the bloodied evidence must be inadmissible." Hey, no murder weapon, no crime! It was just a lot of misunderstood coincidences. The criminal walks. The defense wins. It is the public and the victim's families that suffer. In the courts and in the politics of America, it is becoming disturbingly clear that winning is all that matters. Controlling the perceptions of the masses is the objective. Controlling information is the key to success. Just ask any dictator of a third world nation. It works, at least for a while.

Don't expect a tyrannical ruler to use the proceeds of blackmail to better the plight of his or her citizens. They, instead, will use their ill-gotten gains to build larger and more deadly weapons (Iran, North Korea, Venezuela, Cuba).

Fortunately in the USA, we enjoy a First Amendment right in our Constitution, granting the freedom of speech. It is a right granted to us by our forefathers that makes living in the United States unique. It allows us the opportunity to seek the truth in whatever form it may take. However, should this right be restricted or eliminated by an overzealous political body, truth will die and a totalitarian regime will ensue. Without the ability to freely exchange ideas, cultures stagnate, people live in fear of dissent, and an overwhelming paranoia takes over what was a very open society. It is these kinds of events that usually end up as bloody coups.

Could this happen in our time? Look for the signs, and you will find the answer. Humanity has yet to change.

But what is it about humanity that makes us so special? We as human beings have always tried to make ourselves the chosen species on this planet. The truth is that we are merely one in a myriad of species to inhabit this planet, and many of us believe that our continued existence remains in doubt. Once again, we are destined to become part of the cycle of life on this planet. That is the truth. To make our existence more than that is likened to believing that the world is flat.

As I mentioned earlier, truth is based on fact. Fact can come in many forms, but regardless of the form it takes, fact will always be a human perception. Overcoming human perception and discovering the ultimate truth will require a journey past human experience and into the essence of life itself.

Chapter Five

Naturally Forming Molecules
The Logic of Matter

Did you ever wonder what other planets in our solar system were made up of? Surely they must be composed of exotic chemicals and rock formations. They must be completely unique and different. Well, as it turns out, research has proven that the planets in our solar system are composed of material exactly like that on Earth. The amounts of certain matter are in different proportions, such as hydrogen, nitrogen, and oxygen, but the molecular structure is in perfect harmony with what we would expect to find on Earth. No surprises. No exotic chemicals. How boring! What was discovered is that molecular structure and the formation process of atoms to molecules is just as predictable on other planets as it is on Earth. However, gravitational differences among the planets would be the largest variable and would constitute variances in rock density and crystalline structure.

In 1953, a graduate student named *Stanley L. Miller* decided to test his theories on the formation of molecules basic to the formation of life. The experiment involved a large enclosed glass bubble with molecules of simple gases such as ammonia, methane, hydrogen, and water vapor. The gases were then subjected to static electric discharges acting as simulated electrical storms. All this was done in an effort to simulate an early Earth atmosphere. After one week of stimulating what was referred to as *primordial soup*, Mr. Miller (now Dr. Miller of UCSD) discovered new molecules. The new molecules formed included simple amino acids, a basic substance common to all living organisms. It is believed by Dr. Miller as well as many of his colleagues that these same processes are occurring on many other planets in the universe. The field of study is known as *exobiology*.

This process has created a great deal of speculation as to the molecular surprises waiting for us on nearby planets. These same elements exist on other planets in our own solar system. Jupiter and Saturn both have dense gaseous atmospheres obscuring their surfaces. These gases consist of ammonia, methane, hydrogen, and water vapor. Sound familiar? Research is currently underway to identify other *prebiotic* planets in our solar system.

The fact that life exists on Earth today is no coincidence. It is not an abnormality of nature. It is nature. Life is the natural progression of molecular interaction. Not only is it true on Earth, but it is also true on countless other planets in the universe. Life is a universal constant. Think of our universe as a field full of seeds. Some will germinate and grow, others will not. When and if a planet produces living creatures depends greatly on the mix of matter left to it by its last manufacturer, the sun.

As I will note in many places in this book, the nature of the universe will repeat itself on many different levels. When matter is produced at the subatomic level, it exhibits spiral decay. This spiral decay, when expanded exponentially, produces spiral galaxies. On a solar level, we observe planetary bodies orbiting around a solar mass. The solar body will retain the greatest energy while the planets, with much less energy, will be spun off during early solar decay. My theory of solar system formation would be something like this: During the early formation of solar bodies, gravitational collapse will cause the formation of heavy matter. Atomic matter will form into nuclei of various sizes, and these nuclei will be stabilized by fields of electrons. The heavy matter, now incapable of maintaining gravitational cohesion with the rapidly collapsing solar body (primarily hydrogen and helium), is thrust free of the center and begins to form a solid body of matter. This solid body is now the early protoplanet. The number of planetary bodies will be proportional to the size of the solar body. Every solar body will be surrounded by its accompanying planetary bodies. These planetary bodies will exhibit orbits in the same direction as the solar body and in slightly different planes from each other. Depending on the forces acting upon the solar body, the orbits of the surrounding planetary bodies will be skewed either above, below, or centered with the central plane of the solar body (not based on the deformation of space-time as is currently theorized, but on the undulating forces acting within galaxies on solar bodies). In other words, the planets are along for the ride, and their orbits will adjust accordingly as the solar body changes its position within its own galaxy. This is the end result of solar formation and decay.

So what does this mean to you and me? It means that for every star in the sky, there are planets in orbit around them. It also means that life is attempting to form in every star system in the universe. Not only is life on Earth commonplace in our universe, but so too are the planets that perpetuate life.

At the time that I wrote the previous statement, the Hubble Space Telescope had just been launched. It wasn't until 1993 that the aberration in the primary mirror was corrected and deep space images could be properly viewed. Along with the spectacular images, slight variations in distant solar orbits could be detected, indicating the presence of orbiting dark matter planetary objects. The term used now by the scientific community to identify these unseen bodies is exoplanets. As of September 2012, 124 multiplanetary systems have been identified. Using spectrographic data from these nearby stars, scientists have concluded that the higher the mass and metallicity of the star, the greater the probability of a multiplanet system.

We, as human beings, have difficulty believing that other forms of life exist beyond our own world. Instead of looking at ourselves as the only living beings in the universe, we should instead look at ourselves as a common example.

It is time that we stop thinking as our forefathers. After all, the world is not flat. We are not alone.

Chapter Six

The Big Bang

Those of you unfamiliar with the name the *big bang* should know that this is the theoretical phrase used by scientists of our time to describe the formation of the universe. It states that "in the beginning," there was a huge gaseous cloud of matter throughout our universe. Eventually, through gravitational attraction of matter, the cloud began to collapse. As the cloud got smaller and smaller, particles of matter accelerated into the center forming an intense and densely packed core—a *singularity*. When the pressures inside the core reached a critical point, the immense body exploded into the surrounding space, forming fragments of celestial clouds, which eventually would become galaxies, nebulae, comets, and asteroids.

To me, this sounds a little too much like Earth being in the center of the universe. Why? Because as I stated earlier, there is no "in the beginning" and there is no center to an infinite body.

I remember reading scientists comments on the expanding-universe theory, and they would always ask the question, "Well, if the universe is expanding, what is it expanding into?" Then, of course, there would be some discussion on what shape the universe would form. Would it turn in on itself? Would it form a *Möbius loop*? That's all fine philosophical discussion, but there are some fundamental problems with the big bang theory right from the start. First of all, time is infinite. The universe was really never created; it has always been here. Just as our bodies have always been a part of the earth, which in turn has always been a part of the universe. Second, infinite bodies, like numbers on a mathematical line, have no center point. The numbers continue to get infinitely larger or infinitely smaller. Therefore, an infinite body cannot have a geographical center in which to collapse into.

So where do we go from here? Well, I have a theory.

Imagine, if you will, the universe as an ocean of energized matter. This energized matter exists at an energy state far above our own. In fact, most of the universe is at this high energy state. (Astronomers currently theorize this unknown substance to be dark energy, although its existence has yet to be discovered.) The energy state is so high that matter does not exist in solid form. Imagine also that this energized matter is not homogeneous. It is heterogeneous. There are variations in the energy state within the universe. These energy states form as a result of the expansion and contraction of space-time. There are higher-energy areas (contracting space-time) and lower-energy areas (expanding space-time). It is at these low-energy areas that things start to happen. A void begins to form in the fabric of space-time. A "black hole," if you will. Around this void, the fabric of space-time pulls in on itself, trying to fill the newly formed hole. However, due to the localized expansion, it is unable to do so. A vortex ensues. As space-time around the vortex swirls, it begins to compress against itself. As it compresses, the fabric of space-time begins to unravel. Constituent parts of space-time begin to separate. This separation causes particles of dark matter to begin to form. Separated, their energy state is lower than the energized matter surrounding it. Their kinetic energy (KE) is reduced.

This is the birth cycle of hydrogen and its component parts of an electron, a proton, and a neutron. During the formation process of a galaxy, an enormous swirl of hydrogen gas is produced. Due to rotational compression, the higher-energy electrons cannot bond with the nucleus of protons and neutrons, so they orbit in shells, passing through the core but unable to reattach to it. As the swirl of matter expands, the outer bands lose kinetic energy. As elements of hydrogen and helium form, they begin to coalesce, forming globs of gaseous clouds that will eventually become stars and nebulae. The swirling matter collides with itself and, over millions of years, forms a series of compression bands. At a far off distance from this event, we can see that the formation has become a spiral galaxy.

At every low-energy area in the universe, we see the same event taking place. From our viewable position, we can see millions and millions of galaxies. Within these galaxies are billions and billions of stars and planets. Additionally, no matter can exist without a black hole. There is no compression of space-time without a void. As a result, the space between galaxies is completely devoid of any matter.

Interestingly, we see very similar events in the body of our own atmosphere. The heating and cooling of our atmosphere causes pressure variations to occur, resulting in the formation of high- and low-pressure areas. Measured in isobars, the intensity of a pressure area is determined by how sharply the pressure falls or rises over a given distance.

Most of us know that low-pressure areas bring weather in the form of thunderstorms and rain showers. At equatorial regions, large low-pressure areas form over the oceans. During certain months of the year, these low-pressure areas form tropical storms. As these tropical storms traverse the ocean, they begin to rotate around the center of the low-pressure area. Depending on the hemisphere, the rotation will either be clockwise (Southern Hemisphere) or counterclockwise (Northern Hemisphere) due to the Coriolis effect. As these storms intensify, gaining energy from the sun and moisture from the ocean, they can typically result in a spiral-formation hurricane. Seen from satellite imagery, the hurricane exhibits properties similar to a spiral galaxy. The formations are perplexingly similar.

In a hurricane, cloud densities and storm activity vary with the amount of energy and condensation present within the storm system. In a spiral galaxy, mass densities and kinetic energy vary within the various solar systems and distance from the core, presently known as a black hole.

In the center of a hurricane, there is no storm activity located within the eye, only empty space. It is the area of lowest barometric pressure.

In the center of a galaxy, there is no mass, only empty space, hence the name black hole. It is the area of lowest energy and is devoid of everything, including space-time.

Without space-time, electromagnetic waves cannot propagate. In other words, no visible light is emitted or can pass through a black hole.

Current thinking about black holes is that they are, in actuality, a supermassive clump of protons and neutrons stripped of electron fields. The gravitational field is so great that not even light can escape. In addition, at the center of every galaxy, intense plumes of excited matter are emitted in bursts of energy perpendicular to the axis of the galaxy. No one seems to know exactly why.

My theory of an "empty space" black hole also holds that the stellar matter at the center of a galaxy possesses the greatest kinetic energy of the solid matter surrounding the black hole. It is at an energy state closest to reforming space-time, or "E" state matter. Black holes are also continuously trying to close themselves and, as a result, set up cycles of expansion and contraction. It is during these expansion and contraction cycles that the formation of high-energy plumes erupts perpendicular to the plane of the galaxy.

As the core stellar matter compresses into the black hole, it is converted into space-time and ejects a plume of high-energy plasma and x-rays. Using the frequency of these emissions, one can determine the diameter of the black hole and, hence, the size or stellar density of the galaxy.

There is another property of galaxies that has not been discussed. That property is galactic magnetic polarity. What this means is that on each side of a galactic disc of stars and planets, the orientation of magnetic fields tends to shift as solar systems pass through the central plane of the galaxy. A plus side (North Pole) and a minus side (South Pole), if you will. As our solar system is swept around the Milky Way galaxy, our sun is continually oscillating up and down in a sine wave pattern above and below the central plane of the Milky Way, dragging our planets along the way.

For those of us on planet Earth, each time our solar system passes through the plane of the Milky Way, there is the potential for our magnetic poles (including that of the sun and planetary bodies) to swap, aligning with the magnetic polarity of our galaxy. This planar passage affects the iron core of our planet, which creates the magnetic fields on Earth. Whether or not the magnetic poles shift depends on the location of surrounding planetary bodies and the degree of magnetic variation of our sun. It is an irregular cycle and occurs on the average every three hundred thousand years. Are we overdue? Unfortunately, the answer is a resounding yes. The next planar passage through the Milky Way will occur at the end of 2012. Should such a shift occur on Earth, the molten center of our planet would follow the magnetic lines of attraction and repulsion and reorient itself with the prevailing magnetic fields of nearby heavenly bodies. This rotation of the molten core would cause the crust and mantle of Earth to be violently shifted, placing tremendous stress on the tectonic plates. The result would be massive earthquakes, tsunamis, and volcanic eruptions of a catastrophic scale. This kind of shift in Earth's crust will also cause piezoelectric discharges throughout the planet, rendering most electronic devices useless.

As Earth's magnetic poles rotate with the sun's magnetic field, our planet will be exposed to high levels of solar radiation normally shielded by our previous North/South orientation. Unfortunately for us, all these events will occur at the peak of the eleven-year solar flare cycle at the end of 2012.

Should we be concerned over such events actually occurring? That depends on what you believe in. Considering the fact that there is nothing that can physically be done to prevent it, my suggestion to you is that you find your way back to God. Continue to read this book, and I will help you to reach that goal.

My greatest concern, following a global catastrophe, is the kind of world that will emerge when the surviving humans rebuild a new society.

We are destined to become the future's ancient race. A millennium from now, people will look back to us for answers just as we now do in our time. The question will be asked, "What did they know, and how did they apply that knowledge? Did they learn from their mistakes, or did they repeat them to their own peril?"

How do you want your legacy to be remembered? Do you even care?

Current theories of a big bang phenomenon are incorrect. The universe is not full of hydrogen gas at the outset of a great explosion as the big bang theory would suggest, but instead, matter is gradually formed as the universe expands locally in some areas, stretching space-time and reducing its energy state, and matter is annihilated as the universe contracts locally in other areas, compressing space-time and increasing its energy state. It is this push-pull effect throughout the universe that explains the seemingly random visible matter density patterns in the universe. It also explains why our Milky Way is passing through background radiation, traveling opposite our direction of travel through the cosmos.

But what is the driving force behind this expansion and contraction? What causes the variations in the energy levels in space-time? Deep space mapping has shown that galactic density patterns form along filaments in space-time and these densities occur approximately 140 parsecs apart. When viewed in three dimensions, the structure appears as a web of interlaced filaments similar to a neural network in living beings. Believe it or not, the same phenomena that drives the structure of our galaxies and stars is the same phenomena that drives the molecular structures within our bodies. It is the same effect but at gargantuan proportions. The effect is known as *Cymatics*. (I will expand on this topic later in this book) Think of our universe as a high energy super fluid, vibrating at the speed of light or its "E" state, and these oscillations interact in three dimensions with one another (three dimensional harmonics). The end result is an elegant distribution of matter and space that pervades the cosmos and provides the dynamic interactive force required to form the physical realm out of seemingly the nothingness of space-time. These visible filaments of countless galaxies form dynamically along harmonic nodes in space-time and, as such, continually expand and contract.

The idea that the entire universe is in expansion is, again, small-scale thinking on the part of our scientists. The universe is not flat. Humanity must come to grips with its myopic view of the universe and accept a grander picture.

So what happens as local space-time expands? Quite simply, more voids are formed, creating a higher density of galaxies, and the observed filaments are created. At a certain point, the expansion will stop, contraction will ensue, and the voids that created the galaxies will begin to close, annihilating all solid matter in the region.

As it happens, the local space-time we inhabit is in an expansion phase. Because of this expansion, we believe that the entire universe is expanding. It skews our perception and results in erroneous conclusions in spectrographic analysis.

Astronomers viewing distant galaxies determine relative movement by performing a spectral analysis of the light emitted from a distant source. It is commonly known that a stellar body that is moving away from your viewable position will exhibit a shift on the electromagnetic spectrum toward the red wavelengths. Conversely, a stellar body moving toward your viewable position will exhibit a shift on the electromagnetic spectrum toward the blue wavelengths. Because of this phenomenon, astronomers have concluded that most of the universe surrounding Earth is moving away from us and the light emitted from distant galaxies is moving toward the red wavelengths. They also believe that the universe is expanding like a balloon, ever increasing in size.

Eventually, it is surmised, the balloon effect of expansion will reverse its course and result in a gigantic collapse. This collapse will begin another cycle of the big bang.

So if space-time is ebbing and flowing, is there a way to know which direction it is traveling around us? The answer is yes. In order to view the direction of space-time, we simply have to note the orientations of the galaxies around us. Typically, a galaxy will form parallel to the movement of the space-time that surrounds it. Hence, our Milky Way disc is traveling like a raft in a river through the cosmos.

We, as human beings, must continue to challenge ourselves to expand our realm of thought. The grand scale of things eludes us, and we find ourselves continually having to rewrite what we thought we knew. So is the progress of science.

(See figure 2, titled "The Expanding Universe?")

Once again, we must force ourselves to decentralize our thinking on time and space. We must weigh the evidence, perceive the facts, and conclude the truth; even if it means giving up our current beliefs.

Figure 2

The Expanding Universe?

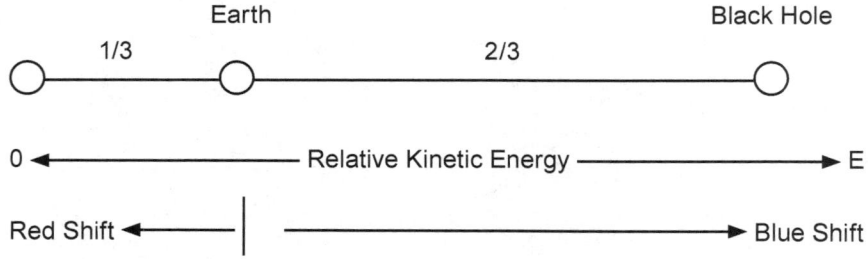

Spectral shift with respect to KE

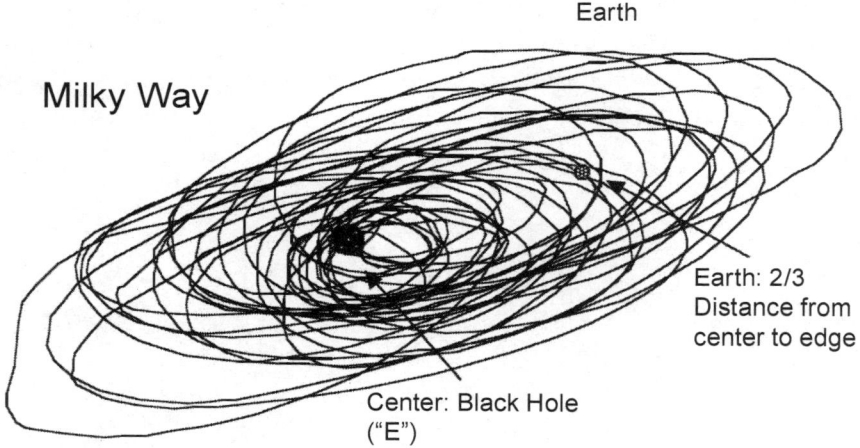

Distant Galaxies

When viewing distant galaxies from Earth, spectral analysis exhibits a red shift. This suggests movement away from our viewpoint (expansion). Or this suggests that our observations of distant galaxies are spectral images of the outer stellar bodies (the outer third). Outer stellar bodies would most likely possess a lower KE than that of planet Earth, creating a red shift irrespective of movement toward or away from us!

(As you can see, my thought process has changed since I created this crude depiction. I am now aware of the dynamic forces which sustain the expansion and contraction of space-time)

Chapter Seven

The Speed of Light

This topic of discussion is going to get me into trouble with a lot of scientists around the world. I will be stepping onto holy ground, and another one of our long-held beliefs.

Albert Einstein proposed in the 1930s some of the most profound concepts on time, space, and energy that the world had ever seen. His formula $E = mc^2$ became the benchmark for atomic physics and the accepted explanation for the motion of particle matter. The letter C in this equation signifies the universal constant for the speed of light. Einstein proposed that matter accelerated to the speed of light squared would become pure energy and that its mass would become infinite.

For many years now, scientists across the world have been accelerating particles of matter in sophisticated linear accelerators, hoping to turn the particles into pure energy. They have been unsuccessful.(Update: The closest we have come to date was the discovery of what is believed to be the Higgs boson on July 4, 2012; a theorized high energy particle previously unrecorded) They have interpreted this as confirmation of Einstein's theories. Now, instead of just trying to accelerate the matter, they have begun smashing particles into one another, creating amazing displays of subatomic frenzy. From the information they have gathered in the accelerators, scientists have begun cataloging the list of ever-expanding subatomic particles. (The current number is twenty-four.)

I suppose all this is worthwhile, but the science community has made a very large error. The speed of light does not exist. Not in the way that you think of your automobile heading down the street. Instead, what I am going to tell you is that the speed of light is actually a time reference based on KE, not a velocity.

Einstein also stated that the movement of objects throughout the universe is relative to your own perspective. If you are sitting in your automobile heading down the street and another automobile approaches you, you would estimate the velocity of the oncoming vehicle to be very great. A person on the side of the street would describe the two automobiles' travel completely different. Each person's perspective would be different and would be relative to their point of reference. This is also true with solar and planetary bodies. Movement of matter in space is all relative to one's viewpoint.

Now, getting back to my statement concerning the "speed" of light, I want to add that light, in and of itself, is not particularly special. Light is an electromagnetic wave. Light also encompasses a wide variety of wavelengths, and is just a small portion of the electromagnetic spectrum.

The only thing that makes light special to us as living creatures is that light waves provide energy to stimulate photoreceptors in our eyes, allowing us to perceive objects. In addition, different species of life are capable of seeing objects at wavelengths that other species cannot.

So why do I say that light does not possess velocity? There are two reasons. First, a photon of light does not possess mass. Second, photons of light travel in waves. This has been established as a scientific fact, and I will not dispute this. What bothers me is that current explanations are far too complex and suggest that photons exist in a form of duality.

Anyone who has thrown a rock into a body of water will notice that the result is the formation of a waveform in the water. The size of the wave is proportional to the mass of the rock. There has been a transfer of energy from the rock to the surface of the water.

Analysis of waveforms in water has been around for years. Floating devices placed in water tanks are observed as waves are formed on the water's surface. The floating devices then exhibit an upward and then downward motion as the wave passes through the water. The distance of the motion of travel is called the amplitude of the wave.

Was there any forward motion of the floating device? No.

How fast did the wave travel through the water? That depends on the temperature of the water. In fact, the velocity of the wave depends upon the temperature of the matter it is traveling through.

How much energy is dissipated at the shoreline? The square root of the distance traveled.

These same principles hold true for sonic waves, as well as electromagnetic waves. So when a rock forms a wave on the surface of the water, the water has not actually moved. Energy is displaced through the water, creating a waveform, and energy is absorbed onto the shoreline. From there, it can be reflected, diffracted, or dissipated.

The general term used as defining wave formation is known as propagation. Propagation of waves requires a medium of transfer, such as a liquid, a solid, or a gas.

The transfer of energy in waveform exists on every scale of matter in our universe. It exists subatomically, atomically, molecularly, and macromolecularly.

When a photon of light is created, it does not travel through space; it merely displaces energy in waveform through space. What's interesting about traveling through space is that it's three-dimensional. Unlike the surface of the water, which is essentially a two-dimensional plane, energy waves in space must exhibit wave motion on multiple planes. The photon does not have multiple personalities; space does. The latest theory on light waves is that they exhibit spiral trajectories. This is a wave form in three-dimensional space and is consistent with multiplanar energy displacement. The time that it takes to arrive at its destination is what I call the universal theory of atomic energy displacement.

On a molecular scale, the rate of that energy displacement is a function of temperature or heat and density. Subatomically, energy displacement is a function of relative kinetic energy (KE rel.).

What does this information do to the meaning of $E = mc^2$?

How does a light wave travel through space if it is not a particle?

I'll get to that in the next chapter.

Make no mistake. Space-time is a medium through which subatomic particles travel and electromagnetic waves propagate through three-dimensional space.

Chapter Eight

Matter and Energy

There is another way in which physicists describe energy on Earth. It is known as kinetic energy. It relates to the motion of solid bodies and is probably the most familiar to most of us. It was described by Newton as "What goes up, must come down." In the simplest terms, kinetic energy is the study of the motion of solid bodies or mass. Solid matter in itself is said to possess mass and, therefore, resist the tendency to be moved. Moving a mass requires energy, or work. That work can either be done through gravitational attraction, or from the stored energy of another form of matter. If another form of matter is used, then it would be described as possessing potential energy.

The source of that energy exists in basically three different forms on Earth. First, it exists as solar radiation; second, it exists as geothermal heat; and third, it exists as gravity, including the gravitational interaction of Earth with the moon. The third method can be observed as tidal activity of the oceans. Realize that solar radiation in and of itself is not a storage medium. What becomes a storage medium are the effects of solar radiation on Earth. Plant life, wind and storm activity, and thermal warming of the oceans and seas are the storage mediums of solar radiation. What is important to note is that this is the total amount of energy available to us as living beings on this planet.

What about petroleum? The petroleum reserves we are using now are in reality the biomass of previous life on this planet. But how is this possible? My belief is that the story of Noah's ark (or something like it) really occurred and probably many times over the millennia. In this scenario, there was a tremendous upheaval of Earth's crust, which resulted in the coverage of all existing plant and animal life by countless tons of earth and rock. Under extreme pressure and stress from the mass above it, the plants and animals were reduced to a conglomeration of carbon-chain molecules, also known as crude petroleum oil, coal, and natural gas.

The cycle of magnetic pole shifts in our planet's history could account for the stratified layers of carbon-chain molecules strewn beneath Earth.

The supply of this oil, coal, and gas is limited and will not meet the demands of our future use indefinitely. Current estimates show that petroleum reserves will be exhausted in forty-five years at our current rate of consumption (see the website http://www.fossil.energy.gov/index.html).

The forty-five year estimate that I referenced above is continually being updated as technology improves in geological mapping and extraction techniques used by the oil industry. In addition, improvements in automobile fuel consumption have helped to increase the number of years cars will be able to operate on fossil fuel. The question still remains, how much are you willing to pay for that luxury?

Realize, however, that we continue to escalate our rate of consumption daily. New reserves are being discovered, but they will not meet the demands of an ever-increasing population. To make matters worse, the oil conglomerates are not seeking alternative forms of energy. Instead, they buy up competing technologies, hampering further progress in alternative modes of transportation, home heating, and electricity production/consumption. They see energy-producing technology that is not petroleum based as a threat to their existence. They are correct.

Unfortunately for us as consumers of energy, we will not have enough time to explore new energy alternatives when the oil wells run dry since few or none will exist.

We, as humans, derive the energy to sustain our lifestyles primarily from the potential energy of stored biomass. Today, this includes petroleum, coal, and natural gas. The rest comes from a variety of sources, including hydroelectric (gravitational), nuclear (solar matter derived), and geothermal heat (residual planetary formation energy).

My point to all this discussion on potential energy is that all the energy we are using today will not be available to us in the near future. The total potential energy for use to us as a planet is limited to the amount that can be stored by our plant life, oceans, and seas. I do not include animal and sea life since they do not store energy directly from the sun. They are consumers of both animal and plant life and, therefore, are not efficient storage mediums.

Is there another form available to us? Yes, but its practicality has yet to be determined. I'm not referring to nuclear power as yet there is no safe location on Earth to store the waste products of nuclear fission. The locations now used to store atomic waste are nearing capacity. Recent reports also indicate that the stored waste may be potentially volatile. Unless the waste material can be rendered inert to living beings, it will continue to pose a hazard to us.

Nuclear fusion is promising, but the containment techniques and replenishment of the plasma core procedures have not been adequately solved. For now, nuclear power has only one safe location among living beings, and that is on the sun.

What I am interested in pursuing is the potential of superconductivity. Before I get into this discussion, I am going to have to make a few more enemies in the field of physics.

First of all, electrons do not flow through wire. In fact, electrons don't really exist the way we now imagine them. In the classical model of the atom, the nucleus of matter is surrounded by a field or fields of individual electrons, depending on the size of the atomic nucleus. These electrons exist in distinct groups or fields in distinct distances from the atomic nucleus. Each field of electrons is said to possess a negative charge as opposed to the positive charge of the atomic nucleus. This harmony of positive and negative charges maintains the stability of the atom. The larger the atomic nuclei or element of matter, the larger the number of electron fields. There are a couple of problems. No one can seem to locate individual electrons in an electron field, and electrons tend to jump from one field to another without a discernible path. This has now become the study of quantum mechanics. This lack of a discernible path in the electron jump has caused many physicists to adopt a string theory of the universe. In other words, everything is connected to everything else through these obscure filaments.

My version of atomic structure is significantly different. First, let's discuss the meaning of heat or temperature. *Heat* is a relative term when discussing matter. When an object is said to possess heat or is perceived as hot, it simply means that the molecular movement or atomic vibration of that object is greater than that of its surroundings. The faster the movement or vibration, the hotter the object or matter. The same would hold true for an object perceived as cold, only its molecular movement would be slower than that of its surroundings. I have postulated the following laws as a result of this discussion.

Craig's law no. 1: Heat is purely a function of the motion of atomic nuclei, not the fields surrounding them.

Craig's law no. 2: The amount of vibration of the atomic nuclei determines the temperature of matter.

Craig's law no. 3: Because of varying mass, atomic nuclei vibrate at distinct rates, much like the natural frequency of a tuning fork. Atomic nuclei, therefore, possess harmonic properties. The letter C in Einstein's equation stands for the frequency of matter based on its relative kinetic energy.

Craig's law no. 4: Solid matter resonated to the square of its mass (or atomic number) will be converted to its energy state, also known as its "E" state.

Hence, $E = mc^2$ refers to a conversion of matter to energy not based on the speed of light but on the excitation of the nucleus and subsequent bonding to its electron field. I refer to this phenomenon as the decompression of matter to its "E" state. In current understanding, it becomes space-time.

Craig's law no. 5: The stability of the atomic nucleus is maintained by wave shells of highly energized matter we now call electron fields. This highly energized matter exists in harmonic bands around the solid nucleus and will remain there until overcome by an extreme gravitational collapse, such as the contraction of space-time. The energized fields of matter exist as a function of the mass of the nucleus. They exist at harmonic nodes based on the frequency and mass of the nucleus.

Now, returning to the statement concerning electron "flow," let me say this: Electricity passing through wire is much like the transfer of energy of an electromagnetic wave, only instead of passing the energy through space, it is transferred through the atomic fields of solid matter in waveform. If the amount of energy passed through the fields is too great, then energy will be absorbed into the atomic nuclei, creating vibration and thus heat. The more the heat buildup occurs, the less able the fields can efficiently transfer energy. Now we get to the heart of superconductivity.

What is superconductivity? Superconductivity is a phenomenon whereby electrical energy passes through solid matter without the buildup of heat. A physicist might describe this transfer of energy as a series of perfectly elastic collisions through the electron fields. The amount of energy that can be passed through such material is far beyond any standard metals in use today. In fact, the best conductors are turning out to be crystalline structures such as ceramics and diamonds.

The ability for a superconducting material to produce intense electromagnetic fields and store vast amounts of energy has already been demonstrated.

There is another phenomenon that has been recently demonstrated by a superconducting electromagnet. It is called the *Meissner* effect. In this demonstration, a solid magnet was placed above a superconducting electromagnetic field. As the field of the electromagnet was increased, the solid magnet rose above the superconducting field and appeared to float above it. The announcement made by the news media was that the superconducting electromagnet had somehow dispelled the magnetic qualities of the solid magnet and this is what caused the solid magnet to float.

My belief is somewhat different. I believe that it was not the magnetic field of the solid magnet that was affected but rather the mass of the magnet itself. The distance that the solid magnet rose above the superconducting electromagnet was a function of the square of the field strength. In other words, the mass was displaced as a function of the square of the energy used. Does this sound familiar? Of course it does. It is the inverse relationship of the universal transfer of atomic energy.

Let me pose a hypothetical question to you now. What would happen if a self-contained superconducting electromagnet were placed on the surface of Earth and its field was sufficient to displace its mass? Would it float? Absolutely. It would exhibit flight characteristics unlike anything we have ever seen. Or have we? This is starting to sound suspiciously like a levitation concept for unidentified flying objects, as it should.

But how would an object like this be able to move through the sky? Maneuverability would take place by altering the polarity of the field. The polarity of the superconducting electromagnet would interact with the fixed magnetic field of Earth and thus provide a trajectory. Altitude control would take place by altering electromagnetic field strength.

What other phenomena are associated with UFOs? Most UFO sightings coincide with electrical power outages, radio wave disruptions, and electromechanical machinery failure. One such example would be an automobile stalling for no apparent reason. Each of these examples would occur in the presence of an intense electromagnetic field. What, if anything, would disrupt the electromagnetic field of a UFO? Answer: the electromagnetic pulse of a nuclear explosion or a powerful lightning strike.

What other phenomena are associated with superconductivity? Superconductivity is the foundation of field strength for "super" electromagnets, and it is also the way in which we will resonate matter to the square of its mass.

Believe it or not, the latter was accomplished in the fall of 1943 during an event known as the Philadelphia Experiment. The code name used by the US Navy was Project Rainbow. Several books have been written on this event, and I will simply capsulize the story for you.

When Albert Einstein was working on his unified field theory (incomplete), a great deal of interest was spawned concerning the theory of bending light waves using magnetic fields. Interest in this possibility came from scientists working with the US military.

The United States Navy thought it would be a good idea to place an intense magnetic field around a battleship in an effort to bend light waves around it. This would supposedly have the effect of misplacing the viewable position of the battleship by altering the light reflected from its surfaces. An enemy ship using either visual sight gauges or radar for its weapon controls would be unable to locate the actual position of the US ship. Every time an enemy ship would attempt to fire at the magnetically altered ship, its ordinance would find nothing but empty sea. The physical position of the battleship would be imperceptible, making it nearly impossible to hit. So utilizing theories as yet unproven, the navy went about placing several massive electromagnets aboard a naval destroyer in Philadelphia harbor, the USS *Eldridge*. They were the largest of their kind ever produced and were powered by additional generators on board the ship.

When the electromagnets on board the ship were activated, a strange event took place. Instead of bending the light reflected from the ship's surface, the ship disappeared from view completely. In fact, the ship was said to be sighted in the harbor near Norfolk, Virginia, shortly after commencing the experiment. It reappeared back in its original position approximately fifteen minutes later when the electromagnets were shut down. A second test was performed after realigning the frequency patterns of the electromagnets. The results of this test were more devastating to the ship and its crew. Although the ship was partially visible during this test, the side effects increased dramatically. Many aboard the ship became violently ill, both physically and mentally. It was reported that during the test, many of the crew lost sight of their fellow sailors. After completion of the test, a few unfortunate crewmen were found embedded in the steel bulkheads, melded into the superstructure of the ship. Many of those that survived the ordeal became mentally ill and never recovered from the trauma they experienced. They spent the rest of their lives in mental institutions. As for the USS *Eldridge*, it is currently scheduled for dismantling and classified as scrap.

Our government has never publicly acknowledged that the Philadelphia Experiment ever took place. I'm sure that based on our current understanding of physics, the truth will not be found. It is here in this realm of mystery that I will shed some light.

As you will recall, I postulated some new laws concerning the nature of matter. Laws 3 and 4 stated that atomic nuclei possess a natural frequency and that this frequency can be resonated. The effect of this resonance is a function of the equation $E = mc^2$. When the ship in Philadelphia was fitted with electromagnetic generators, the designers produced a resonating magnetic field. This field coincided with the natural atomic structure of the ship—namely, steel. As the field strength of the electromagnetic generator increased, so did the atomic vibration of the steel. When the resonance of the steel reached its energy conversion state (decompression of matter), the ship, as well as all those within the ship, vanished. Shutting down the generators restored the ship to its original energy state and hence the rematerialization. That the ship changed locations is not surprising since the ship was no longer under the gravitational influences of Earth while energized. The overall question is, when the ship disappeared, where did it go? More importantly, where did the crew members go that would have caused them to go insane?

In the last chapter, I ended the discussion with two open questions, one of which asked, "How does a light wave travel through space if it is not a particle?" The answer is one that was posed in the 1800s when it was discovered that photons exhibited wave characteristics. Back then, it was dubbed as the great aether, or ether. The aether was an odorless, colorless substance that permeated the universe and was the universal transfer medium of electromagnetic waves. In fact, the aether was used to solve a wide variety of physics anomalies that could not be explained any other way. After the Michelson-Morley experiment failed to prove the existence of an aether using angled mirrors and a light source, the idea of a great aether was abandoned.

Why did the experiment fail? Because the medium through which the experiment was conducted (space-time) was flowing in the same direction as the test equipment. Hence, no relativistic movement!

With Einstein came the next step in solving the mysteries of matter and energy—particle physics. The theory of an aether depended on a tangible substance. Particle physics did not.

So here we stand today, relying on space-time theories based on vortexes and quanta and strings. We have space expanding into "something," but we don't know what or for how long. Astrophysicists can't find enough matter in the universe to support their expansion/contraction theories for a big bang to occur. Electrons are jumping and flowing through matter. It's no wonder science has not been able to solve the mysteries of the universe.

What I am going to tell you is that, yes, there is a type of aether permeating our universe. It surrounds us, it envelops us, it is in all places in the universe, it is a part of all matter, it is basic to the understanding of ourselves and our universe, and lastly, it is totally undetectable to any instrumentation. (The last statement may eventually change.)

Proving the existence of this aether is comparable to a fish in the ocean trying to prove the existence of water. The water surrounds it, permeates it, and makes up everything used to explain it. To the fish, the water does not exist; only the solid objects within it exist. The existence of an aether is a real and not imagined truth to our existence. It is the medium through which all electromagnetic energy passes. It is the medium in which the spirit world exists.

Today, physicists have replaced the term *aether* with the term *quintessence*. Sound familiar? Remember that when I described an atom of matter, I described a nucleus of solid matter surrounded by an energized field. The energized field maintains the stability of the nucleus. What is the energized field? If the crew members of the Philadelphia Experiment ship were alive today, they could tell you firsthand. It is, after all, solid matter excited to the square of its natural harmonic frequency. What excites matter to this state naturally? A collapse of space-time.

Only a small portion of our universe is made up of stabilized matter—matter created by our solar birthplace. The rest is unattached, free-roaming energized matter. It is the transfer medium of the universe.

How fast is energy transferred throughout our universe?

What is the speed of light?

This will bring up an interesting phenomenon because speed is a reference to time, and time is not constant. (See figure 3.)

Remember that I stated earlier that mass or matter set into motion possesses kinetic energy. To move or accelerate a mass requires the expenditure of potential energy. In other words, to be able to drive your car to the grocery store to buy milk requires the expenditure of potential energy or fuel. Increasing the kinetic energy of matter has another effect as well. It changes the relative time of the object as well as its inhabitants.

As we pass through the universe on planet Earth, Earth thus possesses a relative amount of kinetic energy as it orbits around the sun. This kinetic energy is very constant to us since there are no large forces acting upon us to either accelerate or decelerate our planet. Because of this, the passage of time on Earth is fairly constant.

Let me use the space shuttle as an example of an object having its kinetic energy changed. To launch the shuttle requires the igniting of vast amounts of liquid hydrogen and oxygen. The effect of transferring this potential energy into motion is twofold. First, it launches the vehicle into space, and second, it increases the kinetic energy of the shuttle. Its velocity around Earth has increased so that its kinetic energy is much greater than it was when resting on the launch pad at Cape Canaveral. In addition, the relative passage of time has also changed. Time passage on the shuttle (as well as its occupants) is slightly slower than time passage on Earth. This has been documented on shuttle missions using extremely accurate chronographs and was originally theorized by Albert Einstein in his writings on relativity.

How does this information affect the universe? For objects with a high relative kinetic energy (KE rel.), the passage of time, and therefore the motion of objects in the universe, will seem to move very slowly. Conversely, for objects with a low KE rel., the passage of time will appear to move very quickly.

Different planets in the universe possess a different kinetic energy; therefore, the passage of time is slightly different for all planetary bodies as well as their inhabitants.

In addition to KE rel., the specific metabolic rate and diurnal cycle of each species also determines its perception of the passage of time.

I am going to take a bit of a side step at this point in the discussion and share a personal story.

When I was eleven years old, my 5th grade math teacher, Mr. Davidson, was beginning a lesson on speed or velocity. He asked his students to give examples of different ways which we measure speed. I thought about it for a minute and raised my hand. "Yes", he said, "What is your example?" I stated, "The time it takes to open your eyes and perceive the world around you." Mr. Davidson, with a pained look of disbelief said, "I don't believe that's the type of example we're looking for." Humiliated, I decided to keep those types of thoughts to myself for many years to come. What I was trying to convey back then, but was unable to do so, was describe our Cognitive Clock. What is a Cognitive Clock? As I stated above, it is a living beings ability to determine the rate at which the world changes about you. Your perception of speed or velocity is linked to the kinetic energy of the world in which you live.

To add a little clarity to this concept let me take you on a short history lesson.

In the 1800's, shortly after the development of photography (Ha Ha), several scientists and inventors began work on moving picture devices. On May 20, 1891, Thomas Edison and William Dickson presented the first public demonstration of a moving picture device to a convention of the National Federation of Women's Clubs which was named, curiously, a Kinetoscope.

During the same time, stereo images were produced using two photographs taken slightly apart from each other and viewed through special lens. The device was called a Stereoscope. Another example of this device was developed in the 1930's and was updated using Kodak color slide film. The images were mounted on a rotating disc which was slipped into a slot in the viewing device and advanced utilizing a mechanical lever. Sold by Fisher Price and still available today, many of us have memories of looking through a device known as a View-Master.

These early breakthroughs in motion picture effects and three dimensional imaging have been adapted by the television and motion picture industries to give us streaming three dimensional television and movies. Due to rapid technological advancements in microchip memory storage, high resolution displays, and increased image refresh rates, we are now able to enjoy realistic three dimensional media from the comfort of our own homes. Future developments of these technologies could theoretically simulate the same effect as the science-fiction movie, "The Matrix". In this particular example, the brain would be fed all of the same stimuli we use to perceive the world around us by providing inputs not just to our eyes and ears, but also to our noses, taste buds, and skin. It would be the ultimate immersive experience and is nearly within our technological grasp today. Think of your own sensory organs as input devices to your mind. The streaming information we receive from the world around us determines what we perceive as linear time.

So, getting back to my childhood school tale, I want to share this. Our relative kinetic energy through the cosmos (KE_{rel}) determines the rate of flow of information to our senses which, processed by the brain, is subsequently engrained onto our soul. This is the Cognitive Clock, and determines our perception of the passage of time and hence all relative velocities of the physical world around us.

Mr. Davidson, I hope this discussion helps to explain my example.

Now let's discuss the speed of light. Realizing that time passes differently for other solar and planetary bodies, how would someone on another planet determine the speed of light? They would probably do it the same way we do it on Earth. Take a light source, reflect it many times over a known distance, and record the time that it took to travel that distance. Simple, right? Not when their clock runs at a different rate from ours, just as it does on an orbiting space shuttle.

Believe it or not, the speed of light will be the same on Earth as it is on other planets. The reason is that the difference in kinetic energy between two planetary bodies offsets the time differences. It is an inverse relationship. The faster you travel, the slower the clock. The slower you travel, the faster the clock. So even though the clocks are recording time at a different rate, the light source will travel at the same speed. That speed being 186,000 miles/second. Many of you familiar with Einstein's work are probably thinking that this is hardly a new revelation, so here's something you can ponder.

Craig's law no. 6: Electromagnetic energy passing through space is limited by the medium through which it passes, as are all waves of various energies. This limit gives us very specific information about the medium itself. The solid matter in which we exist has a very specific kinetic energy, which we can quantify. But the quantification must be related to space-time. This relative kinetic energy (KE rel.) determines how much additional energy must be added to reach an energized state. The energy required to resonate matter to the square of its natural harmonic frequency is determined by its initial relative kinetic energy (KE rel.). The greater the KE rel., the less energy required to reach an energized state. Conversely, the lower the KE rel., the more energy required to reach an energized state.

Craig's law no. 7: The kinetic energy of matter can be increased in a static state. Although linear acceleration of matter increases its kinetic energy, to do so requires excessive amounts of potential energy. Matter cannot be brought to its energy state through linear acceleration (unless you possess a miniature black hole with an infinite gravitational field). Matter can, however, be brought to its energy state through natural harmonic resonance.

On Earth, since our kinetic energy is fairly constant, the rate of energy transfer, and thus the speed of light as we interpret it, is 186,000 miles/second.

Is there a faster way to get through the universe? Yes. A solid object given an initial trajectory vector and excited to an energy state of matter will traverse the known universe virtually instantaneously. I call this as-yet-to-be-discovered phenomenon atomic displacement. It is identical in nature to the electron field jump. There is no boundary at the speed of light when matter makes the transition to its energy state. Space no longer has three dimensions. Distance has no meaning. Time has no meaning. You have reached "E," the omega state, and you are not alone. You have crossed over. You no longer have to believe in the existence of God, for you have now become one with God.

How do you stop once you have converted matter to the energy state? Simple. De-energizing the superconducting electromagnet while in the presence of a space-time vortex (galaxy) recompresses matter, reducing KE, and stabilizes the matter into a solid. You have returned from an "E" state and are back in your own space-time reference.

Now getting back to my earlier statement concerning superconductivity as a possible solution to our energy needs, let me say this: Superconductivity has the potential to solve our energy storage and transportation needs forever. However, Earth can only provide us with a limited amount of energy supplied by our sun. Current experimentation with superconducting materials has yielded some hopeful results. The largest of which is the discovery of high-temperature superconducting material. Even though the material is considered high temperature, it still requires extremely cold temperatures to operate. Large technological strides will need to be made to turn this physics curiosity into practical applications—applications that will take us into the next technological awakening.

What should we do in the meantime? Time is running out on our usage of coal and petroleum products. Along with this usage are the annihilation of our rain forests and the burning of the great amounts of biomass they represent. Combined, they introduce an extremely high level of carbon molecules into our atmosphere. The effects of which are far-reaching and deadly to us as humans. Not only will the carbon molecules increase the greenhouse effect on Earth, but the molecules will also contribute to the formation of clouds by increasing the amount of *condensation nuclei* into our atmosphere.

Depletion of the ozone, a rise in the atmospheric temperature, the loss of oxygen-producing plants, including those of our oceans and seas, and the unchecked rise of global population will all contribute to a giant meltdown of our polar caps. When this happens, the majority of Earth's surface will be covered in water. Storms will wreak havoc on all continents.

Eighty percent of the world's population lives in cities on the coastlines of oceans and seas. They will be the first to go. The rest of us will starve or be left to the mercy of the remaining animal or insect populations.

Does all this sound familiar? Prophets and scholars have anticipated an apocalypse for many centuries now.

Does all this really have to happen? No. We can make the conscious decision now to begin relying on the natural energy provided to us by our sun.

We will need to make major changes in the ways in which we consume and conserve energy. Our homes will need to be redesigned, our cities will need to be dramatically altered architecturally, our method of mass transportation will need to be overhauled, and our automobiles will need to change drastically in their form as well as their function. Current world population will most likely need to be decreased. Our dependence on buried energy reserves needs to be stopped.

We are beginning to see the effects of our industrialization of the planet in almost every ecosystem. It will be a difficult transition for many of us.

Consider the choices, then consider the price. Whatever that choice is, we must begin to make it soon.

Realize that the above-described events will take a considerable amount of time to unfold. The rate at which they do unfold depends on how fast our global population rises. An apocalypse occurs as a result of drastic change in a short period of time. The magnetic pole shift that I described earlier in the text is just such an event. In addition, at the peak of the eleven-year solar cycle in December of this year, we can expect large coronal mass ejections (CMEs), which will expose Earth to very intense bursts of radiation. This is a deadly combination for life on our planet. We will have very little time to prepare for this disaster unless we know that it is coming. Topographical changes, which will occur during this event, will be unpredictable, making it impossible to know where to take shelter. Food and water will be of primary concern. Protect yourselves and your family the best that you can, and work together to survive.

Should these events not occur, heed my warnings that I have listed above and consider yourselves extremely fortunate.

Figure 3

Time

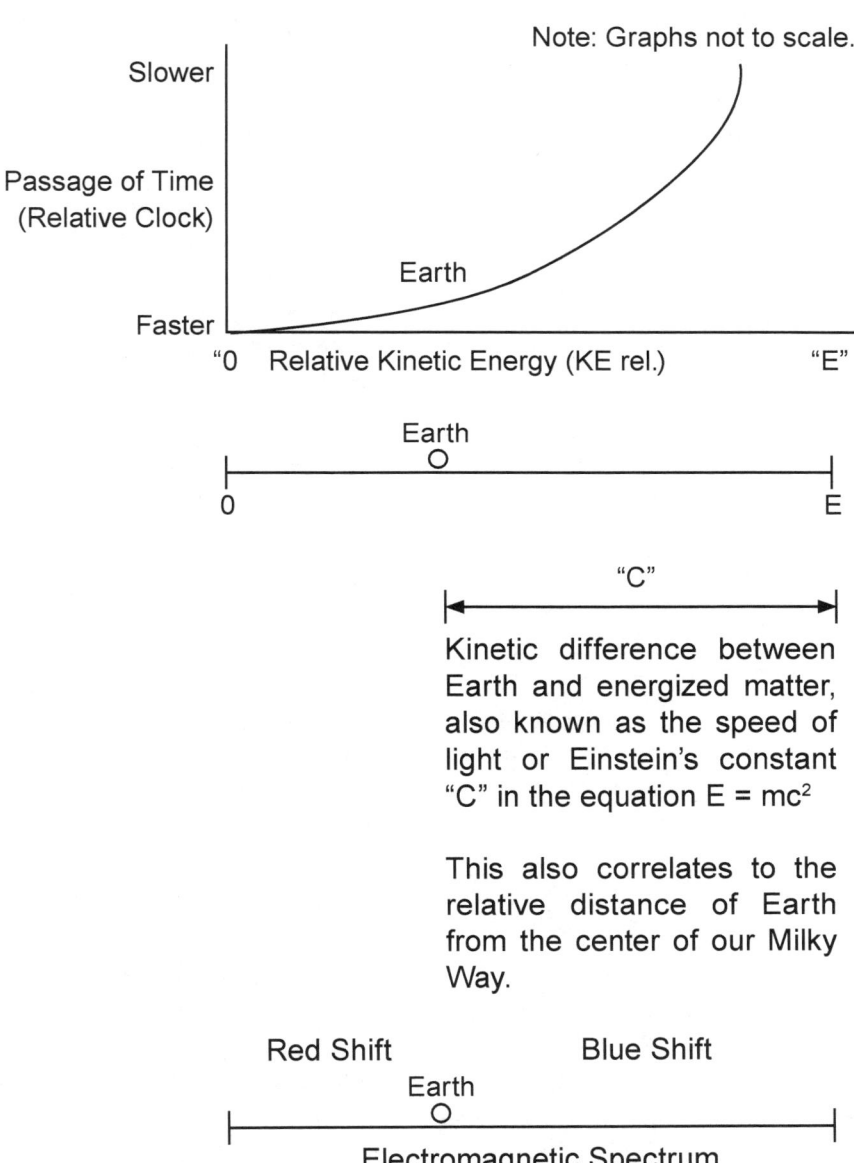

Note: Graphs not to scale.

Slower

Passage of Time
(Relative Clock)

Faster

"0   Relative Kinetic Energy (KE rel.)   "E"

Earth

Earth

0   E

"C"

Kinetic difference between Earth and energized matter, also known as the speed of light or Einstein's constant "C" in the equation $E = mc^2$

This also correlates to the relative distance of Earth from the center of our Milky Way.

Red Shift   Blue Shift

Earth

Electromagnetic Spectrum

Chapter Nine

The Harmonics of Matter

Since I have opened Pandora's box on atomic structure and have described atoms as having harmonic properties, I would like to expand the discussion to include molecules.

The formation of molecules—for example, CO_2 and H_2O—are to be considered fairly straightforward by most chemists. You take a couple of oxygen atoms and combine them with a carbon atom and you have carbon dioxide.

Looking into the periodic table of elements, you will find that the bonding process of the atoms is due to the availability of open electron fields, or valences. The strength of these bonds can be known as covalent, van der Waals, strong, or weak atomic force. The periodic table of elements is constructed to show relative atomic weights (with that of hydrogen) and the amount of free electrons or relative charge. With this information, chemists can postulate or derive chemical composition, commonly known as molecular structure.

I view the periodic table of elements slightly differently, as you may have already guessed. Since I view individual atoms as vibrational in nature, then I also view atomic bonding as harmonic in nature. The relative strength of the molecular bond is then directly related to the harmonic resonance (or dissonance) of the elements in question. The better the match, the more likely the molecule will form on its own. Realize also that elements that do not interact harmonically will not form molecular structures.

During the combination process of atoms or molecular formation, discrete angles form between each atom. The angles between atoms will shift slightly depending on their temperature since each atom will vibrate at a slightly different frequency. This accounts for volumetric differences between solids, liquids, and gases of the same molecular structure. This three-dimensional molecule forms what I would like to call a molecular chord. The musical inference is completely intentional. Linking these chords with one another is a matter of time and level of interaction. Nature acts as its own composer of molecular music. The similarity of molecular harmonics to musical ones is by no means coincidental.

Current science discusses the bonding of elements to one another as a function of valence octets. If one atom has one valence electron (hydrogen) and another has seven valence electrons (chlorine), the resulting bond will combine to produce a hydrogen chloride molecule. The total number of bonded electrons forms an octet.

I would prefer to describe the same process as two atoms seeking to produce a harmonic node based on octave wavelengths. The combination of hydrogen and chlorine atoms is a function of matched harmonics. Octet or octave interaction of atomic fields is fundamental in understanding chemical composition.

Just as a musician would write a melody based on a twelve-note octave scale, so too does nature. Note that harmonious melodies are much more pleasing to the ear than are nonharmonious ones. (Brain-wave activity is directly influenced by musical composition through our auditory senses.) Also, the more beautifully the music is constructed, the more emotionally we respond, just as we would to a beautiful bed of flowers or a forest of trees. As certain notes played on the keyboard of a piano are either in sonic resonance or dissonance, so too are the atoms or elements of matter. (See figure 4.)

As I stated in chapter 5, a vast array of molecules form naturally and consistently throughout our universe. These particular molecules also make up the building blocks of life itself. Looking again at the periodic table of elements, see it now as a keyboard of notes seeking harmony with one another. Studying this phenomenon will bring greater insight into the natural logic of atomic interaction.

The molecules that form so readily in our universe also react (or harmonize) with one another, causing the formation of complex amino acids and, eventually, living organisms. I call this process the symphony of life. I believe that luck and chance have very little to do with this process. I believe that the physical universe serves a very specific purpose and that its existence is not happenchance.

It is time to discuss the big picture. The big picture is a thing we call God.

Figure 4

The Chord of Nature

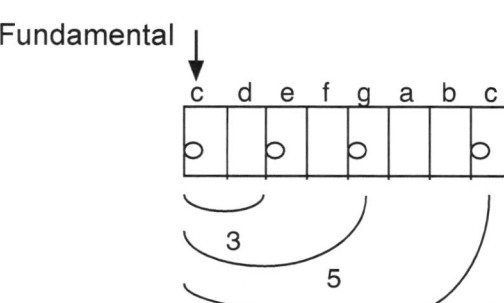

Fundamental

Proposed by Joseph Sauveur in 1701. This chord was supposed to elicit magical tonal qualities to the listener.

These same notes, which make up part of a series of harmonics, repeat over several musical octaves. The first two harmonics are known as a natural third and a natural fifth, respectively.

The Chord of Life

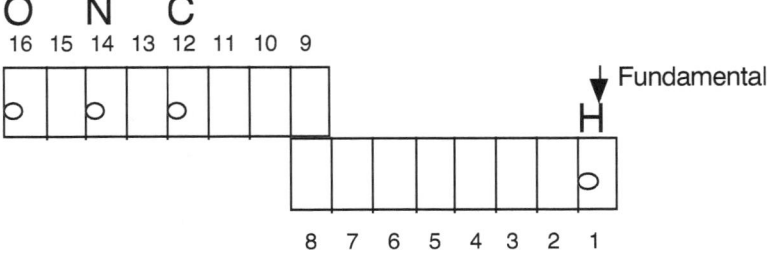

The fundamental elements of life are oxygen, nitrogen, carbon, and hydrogen. The numbers on the eight-note scales represent atomic weights. Because of their mass, lower atomic weights vibrate at higher frequencies than higher atomic weights. Notice that there is a skip in the octaves between hydrogen and the other elements. Realize also that organic molecules form from the least complex and lighter elements on the periodic table of elements.

Chapter Ten

Heaven and Hell

Before I delve into the concept of heaven and hell, I would first like to start with what I believe is the definition of God. As most of us would agree, God is the most supreme being in the universe, maker of heaven and earth, and the one who sits in judgment of our immortal souls. Anyway, that's pretty much the way I remember it from Sunday school.

Over the years, I have come to view God not so much as a supreme individual but as a conglomeration of all living things. In other words, I view God as the influence behind life itself. Not only do I view God as the force behind life, but I find God in all things. God is an integral part of matter itself. After all, living things are composed of matter just like rocks and streams and air. Don't get me wrong; I absolutely do not believe that these particular items that surround us are alive. They are not. However, they do possess the potential to become part of living organisms. Therefore, God is in all things.

I also believe that each of us possesses a soul. I imagine the soul as an individual entity, bound so, as the result of a living organism we call self. Unbound, we know it as spirit. I view all unbound spirits as the body of God. I also believe that there is an integral link between the soul and the spirit. Both are the same entity. One is bound; the other is unbound.

Conscious thought, as defined by the science community, is believed to be a function of brain wave activity. This brain wave activity is typically represented as patterns on an electroencephalograph or EEG. What the science community cannot tell you is what force drives these patterns within our brains. In advanced living organisms, brains serve to provide two primary functions, namely, voluntary and involuntary responses. The involuntary responses are known as autonomic functions, including the cardio vascular function and pulmonary function or breathing. These functions serve to keep our bodies alive without voluntary effort. Voluntary function is what we would define as conscious thought or the ability to choose our relationship with the physical world around us. When a person loses consciousness, such as in a deep coma, most advanced brain wave activity is lost. (Yes. Elvis has left the building) Why? Neurologists will probably have multiple rational explanations based on electrochemical processes within an inactive brain or torpor, but I will give you the true answer, Physical bodies are essentially biological robots; exquisite machines orchestrated by billions of atoms in perfect synchronicity. The soul is the driving force of conscious life. The soul is the oscillator in a biochemical brain. Just as a mechanical robot requires a CPU and an active program to operate, biological robots need a spiritual force to drive brain wave activity and perform voluntary activity. As we are beginning to discover, this brain wave activity is three dimensional in nature. Memories are essentially holographic. This holographic activity within our brains is intrinsically tied to the spiritual realm. Getting the scientific community to accept this fact will take some time to accomplish.

In chapter 8, I mentioned that a great aether exists throughout our universe. I also mentioned that it is in this aether or energized matter that the spirit world also exists. It is the domain of God. Note also that energized matter does not exist in any one point in space. Remember that electrons that surround atomic nuclei cannot be located. Scientists refer to electron fields as shells because electrons can be at any point on the shell at any given time. Therefore, let it be known that just as the energized matter in the universe can be in any place at any time, so also can God. In fact, isn't this the way in which we imagine God to begin with? Isn't this an integral part of the faith in God? God's existence is not just a matter of faith to me; God's existence is a matter of truth and reality to me.

Praying to a higher power is, in fact, seeking guidance and strength from an entity figuratively and literally at a higher energy state than we exist. How did our ancestors know, or is this just coincidence?

Now, I would like to examine the concepts of good and evil. *Webster's Collegiate Dictionary* defines the noun *good* as "something conforming to the moral order of the universe" or "praiseworthy character" and the noun *evil* as "something that brings sorrow, distress, or calamity" or "the fact of suffering, misfortune, and wrongdoing." One other definition I would like to quote is for the noun *moral*, "of or relating to principles of right and wrong behavior." Note that in each of the definitions of *good* and *evil*, the key element or pattern is one of behavior and not one of spiritual guidance. In other words, "the devil made me do it" mentality of right and wrong.

Behavior that is beneficial to others is good. Behavior that is detrimental to others is evil. Are both of these learned behaviors? Not necessarily. Unfortunately for us as living things, we have to deal with the realities of survival. In order to continue as a species, humankind must make decisions that affect the lives of other living organisms. Know that this behavior is similar to the behaviors of all other conscious species on the planet. I have, of course, left out plant life since plants don't consciously make decisions to consume other living things.

I believe that humankind, left to its own accord, would eradicate all other living organisms in order to ensure its own survival. If you don't believe me, then just take a look at the current endangered species list. I suppose this is the right of any dominant species inhabiting the earth, which is, of course, what we as human beings are—the dominant species. We are not the chosen species of the universe. Never forget that our planetary ancestors were reptiles, not mammals. In their time, they were the chosen class.

So let me get back to my discussion on behavior as the driving force of survival. As I mentioned in chapter 1, all living things require the nutrients of organic matter to survive. Organic matter is former living matter. Our decision then is to choose which living matter we will consume to survive.

We, as a species, have done a pretty good job of convincing ourselves that we are the only living organisms on the planet with souls. Therefore, the destruction of other animals to ensure our existence does not leave us with any moral dilemmas. After all, we are the chosen species, right? What harm could it do to destroy another living being if it has no soul?

Religion has helped us to learn to respect other individuals, to avoid confrontation, and to respect the sanctity of human life. I suppose this is to ensure that we don't start viewing one another as "a nice lunch."

There are still many places on our planet that the only religion anyone has known is that of survival. Inner cities, third world nations, and remote jungle tribes are a few examples. The individuals in these examples do not share the same values of good and evil as would an individual brought up with a mainstream religious background. Their primary concern is to survive, no matter what the cost. Human life takes on a more elemental existence as does its value. We consider the lack of respect for human life a cheapening of life itself, yet we consider the lack of respect for other animal life as appropriate behavior. I want to tell you now that I consider the lack of respect given to any living creature an affront to God. I will explain this a little more in a moment.

So now that I have described good and evil behaviors as survival based, let me ask what the appropriate consequence should be regarding punishment. Judaism and Christianity would suggest that one's eternal soul should be banished to hell for eternity for sins against God. Accordingly, only pure or saved souls should be allowed into heaven.

What I am going to tell you is that heaven and hell are not individual places. There is no such thing as absolute good or absolute evil. There is, however, the spirit world and the physical world. Herein lies our confusion. When we die, our souls all go to the same place regardless of how many sins we have committed during our lifetimes.

So I can hear you asking, what is the consequence of our sins? Remember that earlier I described God as a conglomeration of all spirits—a body of spirits. At the time of our death, when our spirit enters the body of God, we are confronted by those spirits with whom we have interacted during our lifetime. They help us to understand the actions and ramifications of our physical life. It is at that moment when the actions of our lifetime are truly scrutinized. It is also at that time when we understand the true meaning of life and receive pure and true insight on God. It is the gift of ultimate knowledge. This is the proverbial judgment of our soul. In reality, the souls that we affected during our lifetime provide our judgment. After all, they make up the body of God. The punishment that we receive is of our own making. We determine the appropriate state of existence our spirit will possess during our next lifetime. This learning process is designed to enlighten our spirit to become a more responsible living being. Note that I did not say human being. All living things make up the body of God. This is the prime basis for respecting all living things.

The animation of inanimate matter is only possible through a spirit of God. Know also that just as the physical world around us is constantly being recycled, so also are our souls. Yes, the soul is truly eternal.

There are some extremely important implications that must be discussed at this time. These involve not only the interaction of humans with other species but also humans with other humans. We are all very much aware of the differences between various races on our planet. Interdispersed among the races are also various religious doctrines. Over the millennia, each of these doctrines has either used or abused the other for some personal justification. In recent times, consider these examples as proof: apartheid of South Africa, ethnic "cleansing" between

Bosnian Serbs and Croats, Jewish Israelites versus the PLO, including various Muslim fanatic groups, Iraq versus Kuwait, ethnic "cleansing" in Rwanda, Irish Catholics versus Protestants, the Ku Klux Klan, and white Nazi supremacist groups.

Conflicts between such countries and groups result in war, famine, disease, pestilence, slavery, or economic collapse. Not a particularly pleasant set of consequences.

What we need to acknowledge now is that our souls have existed in the past as all these various races and religions. Not only has each one of us lived as another race with other religious beliefs, but we have also lived as a man or as a woman. Discrimination against a fellow human being is a failure to acknowledge our own spiritual past. To actively or passively hate or disrespect another is to hate and disrespect ourselves.

Think about your own personal feelings concerning ethnic groups and religious doctrines different from your own. Are you accepting of these differences?

Ponder this, if you will. As little as two hundred years ago, most of mankind was landlocked on the five major continents. Travel between continents was extremely risky, and very few individuals had the resources to do so. Any student of *Charles Darwin* will tell you that any species with little interaction outside its own environment will tend to develop genetic traits specific to its surroundings. In other words, for human beings living in equatorial regions around the planet, skin tone will eventually be genetically altered to take on a darker pigment. In evolutionary terms, this change in skin tone is a genetic safeguard against melanoma or skin cancer. Therefore, each race on Earth has developed as a result of environmental differences on the planet as well as the genetic codes of their ancestors. So as you look at one another on the street, realize that the only difference between you and another human being is your genetic history. No more, no less.

The body is only the vessel through which our soul experiences life. All exterior appearances are superficial in the embodiment of God. Stop sorting yourselves into categories, and start respecting one another as simply another physically bound soul in a world struggling to survive. Animals included! One thousand years ago, the matter that made up your body was no more than a pile of dirt. One thousand years from now, your body will have once again decomposed back to the earth as a pile of dirt. Once you have accepted this fact, you will never look at another living thing in the same way again.

It is God's desire that we love all others, no matter what differences we appear to have. It is acceptable to dislike what a person does to others and to stop them from doing more harm, but we must always respect them as a living soul and love them for that alone. Does this mean forgiveness for nonchalantly taking the life of another? For me, no, it does not. Punishment should always be death; otherwise, we minimize the value of life. This may seem like a harsh penalty and a contradiction to God's will, but there will be a positive consequence. Remember that this spirit will return to us again in another life to make up for previous wrongdoings.

Isn't it interesting that human beings have very little trouble euthanizing an animal that has killed a human but cannot seem to set the same standards of behavior when it comes to humans killing other humans? We will put down a mountain lion, bear, or wild dog, yet will not do the same to a human who commits the same act. This double standard of ethics overestimates the value of human life and underestimates the value of animal life.

What about abortion? Isn't this another form of killing? Yes and no.

But before you end up throwing this book across the room in anger, let me explain why I believe this way.

In order to understand the birth of a child and how the spirit enters the body of an infant, I must first discuss how and why the spirit leaves the body at death. Most of us with a basic belief in God also believe that when we die, our spirit passes into heaven, hell, or purgatory. Regardless of which realm you believe in, the fact remains that you believe in a transference of the soul from the physical body to God. The basis for this belief is that the human or physical body can no longer support our spirit. There is no longer a viable vessel for our spirit to support. The organic structure we call self is no longer self-sustaining. This is what I want you to remember and consider when analyzing and understanding the emergence of our spirit into the physical world—the ability of the organic structure to become self-sustaining.

The human body takes on many forms in the developmental stages of birth. From embryo to zygote to fetus to infant to birth, the human baby is only capable of sustaining life at the latter stages of a pregnancy. When the organic structure of the baby is sufficient to become self-sustaining, then life will continue as body and soul. Remember that all things in the universe are in a state of transition. This includes the transition of spirit to soul. The nourishment that a mother receives during her pregnancy is transformed into living tissue. As the physical development continues, so does the transition of the spirit into the physical world. Conversely, when the physical body withers with age or disease, the soul transitions to the spirit.

Many individuals at the brink of this transition have verbalized their link with the spiritual world. Loved ones of the past, deceased and buried, help to ease the individual's transition with comfort of the knowledge that they are welcome in their new home. Always know that the spirit is eternal. It cannot be destroyed, but it can be postponed from entering the physical world. Understand that terminating a pregnancy in the early stages of development is not the destruction of a spiritual being but the disruption of a biological process sustained by the mother's body. To me, a life has not been taken when a pregnancy is terminated early in the formation process of the fetus. However, once that fetus becomes large enough to support life (a spirit) on its own, then that organism must be allowed to survive and receive God's greatest gift—a soul.

Know that the spirit of the unborn child does not enter the physical realm until the birth event (separation of mother and child). Two spirits do not normally occupy the same physical body during the pregnancy process. Although the physical form of the infant grows within the mother's womb, spiritually the infant does not become a conscious entity until the body and soul unite independently of the mother's body. Evidence for this phenomenon has been noted by young children who can recall their previous and current lives including how they died and how they reemerged back into the physical world. It has been described as a sudden rush or flying into the infant's body from above upon delivery. Psychic Mediums are also aware of these spirits in Heaven, waiting alongside deceased loved ones from whence they will share the birth event with the family.

Termination of a latter-stage (third trimester) normal pregnancy for the monetary gain of clinics, doctors, or for organ retrieval has become a disturbing trend in medical science lately. Greed is not a valid excuse for murder. We need to address this issue with proper criminal legislation. In terms of early-stage (second trimester) normal pregnancies, medical science will have to determine the degree of maturity a fetus has achieved in determining early-intervention procedures and life-sustaining techniques. I hope that physicians remember that quality of life is of equal importance to the sanctity of life. An excessively early intervention may not always be in the best interests of the person who has to live handicapped by pain, discomfort, deformity, or retardation for the rest of their life because an overzealous doctor had something to prove to himself or herself or to others.

Does this mean I favor abortion as a practical form of birth control? No, I do not. I would prefer that any couple engaging in sexual activity do so with the knowledge and foresight of preventing a pregnancy with proper birth control measures. I view abortion as a last-resort measure and do not encourage its use as a practical means of population control. Note also that I insist that mankind limit its practice of unhindered population growth.

Our unwillingness to prevent overpopulation is already causing shortages in water supplies across the globe. Dwindling water supplies mean less drinking water as well as a lack of water for crops, limiting their production, and thus causing starvation. Because of this, there is increasing pressure on nations to find new sources of water. Unfortunately, most of these new sources involve destroying large ecosystems, national parks, and wildlife refuges by flooding river valleys with large dam projects. Not only is this practice prohibitively damaging to the environment, but it also reduces the available water to those living downstream from the dams. We must seek a natural limit to human population before we exhaust our natural resources.

According to the USGS, only 2.5 percent of Earth's water supply is freshwater. That amounts to just over 8.4 million cubic miles of freshwater on Earth. Of the 6 billion plus people on Earth, we are consuming 54 percent of the planet's freshwater supply. By 2050, we will have added 2.7 billion people. That's a 45 percent increase over today's population. If consumption remains unchanged, 85 percent of all freshwater will need to be consumed to sustain humanity, assuming we can even get access to it. This data does not include the amount of polluted freshwater that results from global industrialization. In other words, the world will be out of freshwater.

What about desalinization facilities? Unfortunately, desalinization of saltwater bodies requires large amounts of energy. In a world fighting for natural resources and the inevitable increase in the price of electricity, water may become more valuable than gold, especially when you are dying of thirst.

At present, approximately 300 million people receive water from desalinization plants. All of these people live in coastal cities in countries with vast economic resources, so the extra expense for freshwater is not an issue. Typically, freshwater in these countries is two to three times more expensive than freshwater from sewage treatment plants, underground aquifers, or reservoirs. Using pipelines to move desalinized water into inland cities is not currently feasible and most likely will not be an option for poorer nations.

The Story of Deer Island

Once upon a time, there was a beautiful secluded island located on a majestic blue lake in the middle of the forest.

Once, during an unusually cold winter, the lake froze completely over. The ice had become so thick that the forest animals began to venture out upon its icy surface. One such group was a herd of white-tailed deer. As the deer approached the island in the center of the ice, they began to see a great cluster of bushes exploding in a deep red hue. Upon reaching the island, the deer could see that this red hue was, in actuality, a vast and seemingly limitless growth of berries. The deer sated themselves until they could eat no more. As the weeks passed, the winter ended, and the spring brought back warm, lush breezes. The white-tailed deer had remained.

As the years progressed, the population of the white-tailed deer began to grow. There were no predators on the island, and life was free of stress for the deer. Although the lake had never frozen over again through the winters, the supply of berries maintained the herd for many seasons. Then, one year, a few weeks before the spring blooms began to form, the deer found the berry bushes empty. Needing sustenance, the deer began to eat the buds and branches from the berry bushes. Many of the bushes survived through the summer, but with an ever-increasing number of deer on the island, the number of berries on the island began to dwindle.

The following winter, the berry supply was exhausted by midseason. The deer had no choice but to eat down the remaining bushes.

The summer that followed was disastrous. Some of the deer tried to swim back to the shore from whence they had come, only to drown from sheer exhaustion. Others fought to the death over the seedling sprouts emerging from the soil. The social structure of the herd was in chaos.

By fall, few deer remained alive on the island. The berry bushes that had sustained the herd were no more.

By spring, the population of white-tailed deer, which had thrived on the island for years, had perished.

Ten years passed in the forest until, one winter, the ice on the lake froze completely over. Animals from the forest began to venture out onto the vast expanse of ice. In the distance, a lone, secluded island could be seen. A deep red hue beckoned the animals to its shores.

What is the moral of this story? Never assume that nature will provide for a species that neglects its role within a limited environment. Life, however, will continue on with or without you.

Now I would like to discuss our role as human beings and our interaction with other species on planet Earth. Earlier I mentioned that all living creatures make up the body of God. Therefore, all creatures possess the spirit of God. Plants, animals, single-celled creatures, and bacteria are all biological products of the spirit of God. What differentiates one biological being from another is their level of spiritual awareness. The lower biologically that a living creature exists, the less aware they become of their spiritual being. This does not mean that a lower creature is less likely to be spiritually guided. In actuality, the reverse is true. This is a very important point.

We, as humans, have convinced ourselves that animals that exhibit complex survival techniques are simply utilizing innate instincts to guide them through their lives. In other words, we don't really understand how they do it, so it must be a naturally born behavior. Let me use birds as an example. Obviously, a bird with a brain the size of a walnut shouldn't be able to circumnavigate its way thousands of miles a year around a continent without some kind of magic. Any zoologist will tell you that the neural network of a brain that size could not possibly do the computations necessary to perform such a task. Studies have shown that birds possess a naturally higher level of iron in their brains than humans, possibly suggesting a magnetic direction source, but that does not explain their usage of the same navigational routes. The current answer? Instinct. Perhaps there is another answer.

It is my belief that animals, such as birds, have a closer link to God spiritually than we do as humans. Given that birds are not as biologically advanced as we are, their brains are more in tune spiritually than ours. Although less aware of the source of their guidance, they still are able to utilize it to their benefit. The fact that birds migrate in huge numbers may also be the result of spiritual interaction among the flock. A pooling of their knowledge, if you will. Others who have studied this phenomenon call the force morphic fields.

We, as humans, have had to develop verbal and written language to communicate because of our large and complex brains. We pride ourselves on our imaginative use of symbols and standards so as to pass on information to one another in a way that no other living creature on the face of this planet is capable of comprehending. Perhaps the truth is, other creatures haven't needed to. Theirs is a simpler and more basic form of communication. It is one that we should be very envious, not vice versa. Realize also that human beings exhibit a very low level of instinctual behavior. I believe that this is a direct indication of our inability to interact with one another on a spiritual level.

So getting back to our role as the dominant species on planet Earth, what should be our responsibility to the rest of the inhabitants?

How about respect?

How about tenderness?

How about the right to life?

How about love?

How about the same things that we demand of one another?

Where then do we draw the line when it comes to our own survival? First of all, we have the right to defend ourselves from any creature who threatens our personal survival. This covers a wide range of organisms, including bacteria, insects, predatory mammals, and of course, one another. How we go about doing this should be done with compassion and with minimal destructive impact.

Another factor in survival is the consumption of organic sustenance. Remember that earlier I stated that all living creatures live off the organic material of other living creatures. Lest we forget, plants provide most of what humans require to satisfy our nutritional needs. They receive their nutrients from the soil, not from conscious living beings. They are the most efficient processors of the sun's energy into a food source. They are readily available. They do not promote clogged arteries, heart disease, or the transmission of infectious viral and bacteriological diseases. They are renewable at a much faster rate than are animal meats. And best of all, they are the least destructive environmentally and spiritually to our planet.

Those who choose to continue to consume animal flesh as a food source must abide by the following message: Hold in reverence, respect, and remembrance of those creatures who gave their lives so that yours will continue and flourish. Never consume more than your body requires (gluttony), and never waste any portion that can be consumed.

Am I advocating vegetarianism as a means of survival? Absolutely. Not only for our nutritional needs, but also as a source of energy for the production of electricity. Plants are easily broken down into flammable gases, such as methane, ethane, butane, and propane, and flammable liquids, such as alcohol.

Research now shows that our dietary shift to more animal protein and processed sugars as a source of nutrition has resulted in autoimmune system regression. Once certain thresholds are met, the human body is no longer able to combat carcinogens in the food supply. The result has been an increasing percentage of cancer, diabetes, and heart disease throughout countries who have adopted diets high in animal proteins and fat and low in antioxidants. Reversing dietary needs to the body to stimulate autoimmune responses decreases the rate of aging and sheds the body of unnecessary weight. Besides, we can just as easily survive on the byproducts of animals, such as milk and eggs, to obtain the protein that our bodies require as opposed to consuming their flesh.

Kill your livestock and live for a few days,
or feed your livestock and live for a few
years.

So now that I've gotten completely off track on my discussion on heaven and hell, let me circle back and discuss an individual we all seem to fear named Satan. Before I reveal who I believe Satan is, let me first describe who I believe you think he is.

I believe most individuals believe Satan is a fallen angel from heaven who is here on Earth to dissuade you from your pursuit of godly deeds. Satan is temptation of all things not of God. Satan is the evil influence in our lives that causes us to sin against our fellow human beings. Satan is in pursuit of our immortal souls, and no one is immune from his influence. Satan is the Antichrist.

This is the way that Satan is perceived today, and this is the way that Satan was perceived in the ancient world. Is this still an accurate description? To me, it is not. I have spent a great deal of time in this book describing a universe that is made up of two things, solid matter and energized matter. The interaction of the two matters makes up the known universe and continues to create new planets and stars. On these planets begins the cycle of life that we are so intimately familiar with on Earth.

I believe that Satan is to solid matter what God is to energized matter. One is the alpha; the other, the omega. The two are integrally linked, and both are misunderstood. It is true that Satan is the master of the physical world. To me, Satan is the physical world. All temptations to possess earthly things are considered to be the work of Satan. Remember the Ten Commandments. Are these not the words of God?

> Thou shalt have no other gods before me.
> Thou shalt not make unto thee any graven image,
> or any likeness of anything that is in heaven
> above, or that is in the earth beneath, or that is in
> the water under the earth.
> Thou shalt not bow down thyself to them, nor
> serve them.

Thou shalt not take the name of the Lord thy God in vain; for the Lord will not hold him guiltless that take his name in vain.
Remember the sabbath day, to keep it holy.
Honour thy father and thy mother; that thy days may be long upon the land which the Lord thy God givith thee.
Thou shalt not kill.
Thou shalt not commit adultery.
Thou shalt not steal.
Thou shalt not bear false witness against thy neighbor.
Thou shalt not covet thy neighbor's house, thou shalt not covet thy neighbor's wife, nor his manservant, nor his maidservant, nor his ox, nor his ass, nor anything that is thy neighbor's.

These commandments were written to ensure peaceful coexistence among mankind. They identify those traits in humans that create conflict among ourselves. Many deal with the obsessions of others or of their belongings. Observe closely and you will see similar behaviors in other species on our planet. We all do what we need to do to stay alive.

Maslow's hierarchy of needs prioritizes those physical things in our life that we require to survive. They include food, shelter, and clothing.

Beyond that lies a deeper truth—the truth of our souls.

Take the next step in understanding yourself and the universe around you. Realize that God is the true source of all life. Satan is the manifestation of the physical world. Worship Satan and you might as well be worshiping a pile of dirt. After all, dirt is just a temporary state of matter waiting to become something else. What that something else becomes can only be determined by God. Is Satan evil? No. Behavior that dwells on the accumulation of physical things or the control of other living creatures is evil. Behavior that disregards the spirit of another living creature by nonchalantly taking its life is evil. Thou shalt not murder.

The physical world exists so that we may find beauty in God's creations. Without a physical world, there would be no place for our spirits to inhabit. The universe would be meaningless. Who would enjoy its grandeur?

Physical life is a challenge on many levels. The highest level is the understanding that physical possessions are not really possessions at all. They are just fancy piles of dirt that we take pride in accumulating. They give us false comfort and security.

What is the meaning of life? Seek the truth in all things, and live by those truths. We will all find a different level in our search, but never stop trying. This is God's will.

Realize also that no one individual or group of individuals has all the answers to your questions. Humanity will only survive if we look to the wisdom of all races, religions, and species on this planet in our search for truth. Perhaps this is a basic truth in itself. I hope that your journey is enlightening. Enjoy your gift of life, and share that joy by helping others find their way to God.

Chapter Eleven

The UFO Phenomenon

I'm sure that most of us are familiar with the acronym UFO. If you're not, then let me define it for you as unidentified flying object. Over the years, and specifically in recent times with the advent of personal video cameras, documentation of UFOs has continued to escalate across the globe. It is becoming more and more difficult to explain away these seemingly impossible videos and photographs of moving objects and points of light. The individuals who catch these unusual occurrences on film, in many cases, aren't even in search of anything out of the ordinary. They just happen to be in the right place at the right time. Speculation as to their motives for creating such documentation is reduced considerably. They are just as surprised as we are.

Before I get into explaining how I believe that UFOs operate, I would first like to shed some light on what I believe to be a motive for an extraterrestrial to visit Earth. Although many individuals believe that Earth has been visited by extraterrestrials for many millennia, I believe that what appears to be an escalating number of sightings is directly related to humankind's development of nuclear weapons. I do, in fact, believe that Earth has been visited many times by entities outside our solar system.

My tie into nuclear weapons as a motive comes from a well-documented incident at Roswell, New Mexico. This incident, which occurred in July of 1947, suggests that an alien UFO crashed into the Mojave Desert and was recovered by the US Army Air Corps. Along with the recovered wreckage were also the inhabitants of the object, some of which were still alive. Unable to keep these inhabitants alive, the US Army called in a team of military pathologists to determine the cause of death and examine the body's physiology. Films of this autopsy were recently discovered and released, but their authenticity still remains in doubt. The location of the remains of the vessel and its inhabitants continues to be a mystery today. The military fervently denies the incident ever took place. They instead claim the object in question was a downed weather balloon, not a UFO.

I prefer to believe that something significant did happen in that desert. Consider this as a possible scenario. Top secret surveillance of Soviet nuclear testing was being conducted using high-altitude balloons and electronic sensors. These balloons were constructed of material with high reflectivity and operated at very high altitudes. A UFO, investigating the area due to recent nuclear weapons testing, observes a bright metallic object floating in the skies of New Mexico. The UFO decides to get a closer view of the floating object. Unfortunately for the UFO, a lightning strike, from developing storms, downs both the UFO and the surveillance balloon. As a result, two separate crash sites exist. One from the remains of the balloon and the other from the disabled UFO. Both sites were "sanitized" by military personnel. The remains of both sites were taken to an air base in Fort Worth, Texas, for analysis.

This was the beginning of the Cold War, and paranoia was rampant. We had been at war for many years and were spying on the Russians to determine their threat to us. For all we knew, this object could have been a secret weapon that had penetrated our defenses. The confiscation of the remains by the military was to verify this threat. The hardware that was recovered was not built by the Russians; it was built somewhere else. I believe we have been kept from discovering where that somewhere else is.

So my question is, what was so important around Roswell that an alien UFO would travel several light-years to investigate? The answer was nuclear weapons testing. Above-ground nuclear tests had been taking place not far from Roswell in a top-secret facility near Alamogordo, developed at what is presently known as Los Alamos National Labs. The first detonation of a nuclear device took place at a location known as the Trinity Site. It was a very "dirty" explosion, releasing a great deal of radioactivity. It was not the radioactivity that got the attention of our alien guests.

Associated with the radioactivity of a nuclear blast is also a phenomenon known as an EMP, or electromagnetic pulse. It is created by the sudden expansion of nuclear particles in the center of the blast that in turn sends out an intense waveform across the electromagnetic spectrum. This is also accompanied by extremely powerful shock waves through the atmosphere, creating a large "boom." Massive amounts of heat spread out from the rapid collisions of atomic nuclei with the molecules in our atmosphere, vaporizing all surrounding organic material.

So which one of these events would an alien find interesting? The EMP. As it happens, an EMP has a very distinctive signature in outer space. This pulse of energy released into space might as well be a calling card to anyone listening with a radio telescope. When we detonated nuclear devices on the surface of the earth, we unsuspectingly announced to the cosmos our technological prowess of destruction. We announced that mankind had ascended from the Stone Age. Wouldn't this type of event cause curiosity from an advanced extraterrestrial civilization? I believe it would and still does. I also believe that a few years after the Roswell incident, someone in a high place in national security recommended to our political leadership that we curtail above-ground nuclear testing and instead detonate nuclear warheads deep below the surface of the earth. Publicly, they could state that this type of testing would be much safer to the population since it would greatly reduce the amounts of nuclear radiation deposited into the atmosphere. This, of course, is true.

But what about irradiation of the aquifers deep below the surface of the earth, which supply our irrigation and drinking water supplies? This is not protected. Irradiated water would account for higher incidence of all types of cancer in human beings. Individuals living in parts of Nevada have a much higher incidence of cancer than the rest of the country. This is partly due to the nuclear waste storage facilities located in this state. Perhaps this is why the American Cancer Society is having such difficulty identifying carcinogens in our food supply.

As it happens, rumors of a cure for cancer in the 1940s were said to involve a process of cleansing drinking water by neutralizing its ionic properties. Unfortunately, this process was and still is not accepted as a viable therapy for cancer by physicians in the United States.

Consider also the effects of suboceanic detonations and "isolated" island detonations, and we have now irradiated the breadbasket of the world's food supply in the heart of the Pacific Ocean. Add to this our underground storage facilities for nuclear waste, and I think you can see the insanity of what we are doing for the sake of national defense.

So what is the reason for underground detonation of nuclear devices? In my opinion, it is to shield the EMP from being transmitted into space. It also makes it very difficult for our adversaries to determine the type of nuclear testing being done, not only on Earth, but also in space. Coincidence? I don't believe so. Why then did the United States admonish the French government for its recent above-ground nuclear testing in the Indian Ocean?

The French government exploded a total of six nuclear devices in the years of 1995 to 1996 in the South Pacific atolls of Mururoa and Fangataufa. Shortly thereafter, due to global criticism, French president Jacques Chirac announced that France would terminate its nuclear weapons testing program. (See, I did write this a long time ago.)

In 1976, the federal government of the United States authorized funding through NASA and JPL for a project known as SETI. This acronym stands for the search for extraterrestrial intelligence. The project authorized the development of large radio telescopes and wide-range electronic scanners, whose purpose was to monitor an extremely wide range of electromagnetic bandwidths. These scanners also have the ability to ignore background noise commonly associated with stars and galaxies. Their job was to find anomalous patterns not associated with known stellar phenomenon. They were and still are looking for electromagnetic pulses originating from a dark matter source. *Dark matter* in astronomical terms is commonly referred to as asteroids, comets, and **planets**. No, this was not a typographical error. Serious scientific research is now being done to find intelligent life on other planets outside our own solar system (*exobiotics*). Unfortunately, in 1990, funding was eliminated from the SETI project due to budget cuts at NASA. Up to that point, no solid evidence had been found to support the research. SETI continues to operate as an entity outside of NASA but relies solely on private funding. So far, we are still waiting for an extraterrestrial signal. It appears that either no one in our universe is setting off nuclear bombs against their fellow creatures, or they simply have reached a state in their existence where they don't need to, or they succeeded and there is no one left to tell us about it. Whatever the reason, I hope that we will continue to monitor the heavens for signs of intelligent life. With the exception of the Hubble Space Telescope, the only resource available to continue this research comes from private donations and the scientists themselves.

The Hubble Space Telescope will end its final mission sometime in 2013. Over the next few years, it will be deorbited where it will burn up during reentry into Earth's atmosphere. In 2018 (or later, depending on funding), the James Webb Space Telescope will be launched into orbit.

Now that I have presented a case for extraterrestrials wanting to visit Earth, how then would they be able to cross vast distances of space to come here? And if they have visited our planet already, why haven't our advanced early-warning radars been able to pick up their targets?

In chapter 8, I discussed the transference of matter to energy by relating it to Einstein's equation $E = mc^2$. I also listed some phenomena associated with UFO activity, such as power outages and failures of electromechanical devices. All these phenomena would occur in the presence of an intense electromagnetic field. I have always found it intriguing that many individuals reporting UFO sightings have described the flying objects as disc shaped or oval. Many have also included that the objects tended to glow and were extremely quiet. The reports also include that the objects hovered, moved abruptly, and then accelerated away at incredible speed or just completely vanished.

Seem impossible? Not to me. Let me describe to you how to build a UFO:

1. Construct a disc of superconducting material large enough so that the interior space in the center of the disc will accommodate a generator, fuel, life-support equipment, navigational hardware, and passengers.

2. Insulate the disc electromagnetically from the center by using a nonferrous or nonmagnetic material, such as aluminum, ceramic, or plastic.

3. The resonant frequency to energize the electromagnet will be different from that of the interior structure. This ensures that the occupants are not resonated with the outer structure. The occupants or passengers will be surrounded by a resonant bubble. The interior will remain unchanged while the exterior is excited to its energy state.

4. The outer structure, or hull, of the vehicle must protect the occupants from the extremes of space and atmospheric differences associated with other planets. The shape of this outer skin must aid in the aerodynamic stresses of traveling through various-density atmospheres while in a solid state. A disc or oval shape would work nicely since there is no front or rear of the craft. Also, since directional travel is simply a matter of controlling polarity of the field with that of the dark matter (Earth), an object aerodynamically similar on all sides would perform optimally in any direction.

5. Nonconductive landing supports must be highly heat resistant since penetration of the superconducting field for landing purposes will cause an immediate excitation of the molecular structure of the landing supports, inducing heat into the members. The result of this heat buildup will burn or char any combustible material in the landing area.

6. Excitation or resonance of the outer hull accelerates the nuclei (increasing their

KE) so that reattachment to their respective electrons occurs. This process decompresses the matter of the hull to its "E" state, or space-time condition, while still operating within a black hole vortex (galaxy).

7. Navigation of the craft is accomplished through a neural biological interface between the vehicle's resonant generators and the pilot in command. As strange as this may seem, the craft is spiritually guided through space-time based on the pilot's mental inputs.

8. The power used to operate the craft is generated by a small nuclear generator using element 115 (discovered in 2004 at the Joint Institute for Nuclear Research (JIINR) in Dubna, Russia). Heat from the reactor core is used to power a thermionic generator. The thermionic generator produces the electricity necessary to power the superconducting electromagnets, which are responsible for levitation above Earth's magnetic fields and also the resonance of the outer hull for space-time conversion (1.21 gigawatts! Sorry, I had a *Back to the Future* flashback).

The motion of these objects will be unlike anything we have ever seen. Why is this? Objects that fly through our atmosphere rely on principles defined by Bernoulli. Bernoulli formulated equations that deal with the motion of fluids and gases. The principle parameters are pressure, density, and velocity. Because of his work, we now understand how birds are able to fly.

Bernoulli's formulas explained the lift associated with an airfoil. The development of the airfoil in aviation has progressed from wing to propeller to turbine fan blades. All these are examples of an airfoil, and all require velocity through a gas to perform work that can be translated into motion or lift.

A UFO or electromagnetic lifting body (ELBO) does not rely on gases or an atmosphere. This explains the ability of an ELBO to hover and randomly change direction, but what about the incredible acceleration of these bodies associated with UFO sightings? That deals with increasing the static kinetic energy of the object through electromagnetic resonance. By increasing the static kinetic energy of the object to just below "E" state conversion, you effectively alter its time reference with respect to ours. This adds a whole new twist to relativity. Remember the USS *Eldridge*!

As I mentioned earlier, we on Earth view time passage as relatively constant since the kinetic energy of Earth is relatively constant. As an object passes us with a high kinetic energy, we would describe that object as having a high velocity relative to our own. We also know that a rapid increase in linear acceleration causes large stresses to be induced into not only the object being accelerated but also its inhabitants. An example would be a fighter jet aircraft being catapulted off the deck of an aircraft carrier, a top fuel dragster accelerating through the quarter mile, or the launch of the space shuttle.

Static increases in kinetic energy are just that. There are no undue stresses to the vehicle or the occupants of such a vehicle. The appearance of great velocity is only the perceived difference in kinetic energy. The great acceleration we perceive a UFO is exhibiting is due to the time reference change between it and us. In other words, our timeframe is not the same as the timeframe of the inhabitants of the UFO.

Why do UFOs glow? To explain this, let me use our sun as an example. As we know from spectrographic analysis, the sun is made up primarily of hydrogen and helium gases. The reactions of these two elements occur as a result of both nuclear fission and nuclear fusion. The excitation of these gases on the sun's surface creates electromagnetic waves. These waves radiate in all directions and at many different wavelengths. Solar flares and sunspot activities disrupt the fairly uniform discharge of electromagnetic waves and have a direct impact on radio signals transmitted and received on Earth. Since the sun discharges a wide range of electromagnetic waves, including those in the visible light spectrum, we would say the sun glows. Part of that radiated energy is used to stimulate the photoreceptor cells in our eyes. When the outer shell of a UFO is energized, it too radiates electromagnetic waves in the visible light spectrum. Just as the sun radiates electromagnetic energy in wavelengths outside our view, so also do UFOs. Do not confuse this with nuclear fission or fusion; it is neither. However, it is matter increased to a higher energy state. This is why UFOs glow.

Why don't UFOs show up on air traffic radar? We are finding out that sometimes they do, but most of the time they don't. The reason that a UFO would not show up on radar is due to the fact that radar antennas rotate. The rotation is known as a sweep. The sweep can either be in a complete 360-degree circle or in a small back-and-forth sweep known as a pencil sweep. The first is for large coverage, and antennas take a few seconds to complete a full sweep. The latter is for a small area of coverage, and antennas can make several sweeps per second. Both have specific uses in target recognition and for target acquisition.

New types of radar antennas are known as phased array. These antennas are not parabolic dishes but instead are flat panels and are capable of scanning larger areas than a dish antenna of the same relative size. They are commonly used in military applications.

So the reason that radar does not pick up UFOs is not that they are invisible to radar interrogation; rather, they avoid detection by passing through the atmosphere before the radar can complete a full sweep. The UFOs that are acquired by radar have usually been sighted from the ground first and then verified by a radar observer. Since a UFO has the ability to change direction rapidly, the ability for a radar to track its position is degraded significantly. Our advanced hardware is just too slow.

Do you still think UFOs are mysterious? Probably. I hope that I've helped to reduce the mystery of their flight characteristics and appearance.

One subject I have avoided concerns the navigation of such a vehicle through the vast distances of outer space. My best guess is that galaxies would be identified by their black holes.

Stellar and dark matter would no longer be identifiable in this realm. Getting from point A to point B? Who knows? To tell you the truth, I haven't got a clue as to how this is accomplished. Since most of this discussion concerns technology that we have yet to discover, much less master, I'll have to leave this topic unanswered. After all, what fun would it be if I gave you all the answers?

As you can see from the added paragraphs above, I have improved and updated my understanding of "E" state navigation principles. Mankind has long believed in some form of telekinesis or mind control of objects. We just haven't gotten to a technological level of development or understanding to make it happen. Accepting that there is link to both science and spirituality will be the first step.

Chapter Twelve

Physics' Holy Grail: The Unified Theory

For the past several decades, scientists have been speculating on a theory that would explain all the forces of nature into a single solution. The solution would have to relate the forces of electricity, magnetism, gravity, and light. To date, science has only been able to relate electricity and magnetism mathematically. The jump to the next level of thinking is going to have to encompass a new train of thought by our scientific community. If the unified theory exists (and it does), then some fundamental principals will have to be recognized:

1. The universe exists as a sea of energized matter, only part of which is stabilized into solid matter.

2. Atomic nuclei vibrate at distinct frequencies. These nuclei are surrounded by wave shells of energized matter that exist just below the "E" state. The shells or fields form at distinct distances from the nucleus known as nodes. These nodes represent harmonic convergence of the energized matter. The number of nodes is dependent on the size of the nucleus or atomic mass. The interaction of all solid matter corresponds to resonant bonding of atomic fields (harmonics).

3. Energy is transferred through the fields of matter (stable or unstable) in the form of waves. Energy transferred through the vastness of space is electromagnetic wave energy. Wave transfer of energy is universal throughout all forms of matter.

4. Solid magnets are unique in nature. The matter that makes up a solid magnet has adjustable fields. Adjustable fields can be shaped. A shaped field has magnetic qualities. These qualities are attraction and repulsion. I would call them resonance and dissonance of the atomic fields. A solid magnet has aligned or

phased atomic fields. The greater number of atomic fields that can be aligned, the stronger the magnetic force.

5. Planetary bodies are made up of elements of matter, the vast majority of which are neither phased nor aligned. Nonalignment of the atomic fields means that planetary bodies have relatively weak magnetic fields. The magnetic fields, which do exist, form as a result of molten iron core rotation below the mantle of the planet. Gravitational attraction is a function of resonant interaction with other heavenly bodies' elements of matter. The larger the planetary body, the stronger the attraction becomes. Escaping the pull of gravity on any object is a matter of nullifying this resonant property of attraction (such as a superconducting electromagnetic field).

6. Solid matter traveling through the cosmos creates a bow wave effect in space-time. Any matter located in the path of this bow wave will be affected. An example of this bow wave effect is the orbital path of Mercury in our own solar system. Mercury's orbit around the sun is slightly lower than the orbits of all other surrounding planets. Einstein explained the phenomenon by creating a space-time vortex around the sun and expanded this logic to create his theory on gravity. Science is just now recognizing the existence of gravitational waves but hasn't decided how they are propagated.

7. The relative strength of resonant properties on matter in the universe is as follows: (a) atomic, (b) molecular, (c) magnetic, (d) gravitational. (Atomic resonance being the strongest and gravitational resonance being the weakest.) These are the identical levels on which wave energy is observed in our universe.

8. Energy transferred through atomic fields of matter is known as electricity. Electricity passed around matter of magnetic potential (iron or ferrous metals) will phase or align the fields of that matter, creating a solid magnet. Resistance of alignment results in force applied opposite the flow of electricity (solenoids and electric motors).

9. Visible light is electromagnetic energy. The confusion of light as a particle is due to the three-dimensional qualities

of space, not the electromagnetic waves themselves. Energy passed through space-time is directional. This is an essential property that makes intergalactic space travel possible. When light waves pass through the vastness of space, they encounter physical debris in the form of gases, planets, stars, comets, asteroids, and dust. The presence of this matter has the effect of distorting the path and properties of light waves so as to appear that light waves bend through space. The bow wave effect on space-time around planetary bodies has the illusion of bending light waves. Just as the air passing over a moving airfoil is displaced and then restored, so too is space-time around a solid object swirling in a galactic formation.

10. Matter that becomes an organic structure or life is made up of binary pairs, tetrahedrons, octets, and formations of twelve (there are twelve harmonics within an octave). Consider that three-dimensional objects in space must have length, height, and width. It takes two points in space to form a line (length or binary pair). Add two more points in space and a square can be formed (length × height or quad). Add four more points in space and a cube can be formed (length × height × width or octet). Notice also that a simple three-dimensional cube can be defined numerically as 2 × 2 × 2, or Euclidean geometry, and relies on cube root equations to define three-dimensional objects. Nature follows this same path when designing organic structure in three dimensions. (See figure 5.)

Life is the harmonic structure of key elements in the universe. Those elements are primarily carbon, hydrogen, oxygen, and nitrogen. The simplest natural state of each is a binary pair. Living organisms form around carbon-chain molecules or octets. The organic structure of glucose, the simplest sugar molecule, is $C_6H_{12}O_6$. The human mouth contains thirty-two teeth. Thirty-two is a whole number divisible by eight. It is also two to the fifth power. DNA strands, which make up chromosomes, are binary structures consisting of four elements (C, H, O, N), eight essential amino acids, and twelve nonessential amino acids. DNA strands are also in the shape of a double helix. A helix is spiral shaped. On a subatomic level, light waves or photons travel through space in a spiral trajectory. (Don't forget, the forces of nature repeat themselves on multiple scales.) On a macroscopic scale, the human hand consists of two opposable thumbs (binary pair) and eight fingers (octet or two quads). Human sensory organs such as eyes, ears, and nose or nostrils are also binary pairs that, of course, connect into brain halves. Biologists call these traits in living organisms bisection, or mirror-image structure. Unfortunately, when you cut something in half and compare the similarities, you only get half the picture.

Had early man not counted his thumbs along with his fingers, we would have mathematics based on the number eight instead of the number ten, and perhaps the mysteries of the universe would have unfolded a little sooner.

Discussion continues on whether or not the universe is guided by some unseen force. The term intelligent design has been coined, which suggests the possibility of a grand plan. Well, if you're not sure, then how do you explain your own existence or the existence of millions of other living organisms?

Many in the scientific community are adopting what is known as crystal theory. In crystal theory, organic molecules can form in either left- or right-hand orientations (mirror-image structure, see above). Depending on the orientation, different amino acids can result.

I would prefer to think of organic structure as auto-organization. Auto-organization results as a consequence of a dynamic environmental exchange. Without getting too esoteric, I will tell you that, yes, this is a grand plan and it occurs at every nook and cranny in the universe. As I have stated, the purpose of the universe is to create life.

So, what is the ultimate truth? How do we combine all of these seemingly dissimilar concepts into a knowable reality. It's called three dimensional harmonics. Mathematicians have been combining volumetric equations from geometry with Fourier Series Algorithms to describe organic anatomical structures. The purpose of which is to quantify evolutionary biology. In other words, scientists and mathematicians are trying to reverse engineer organic structure. By combining cube root equations (necessary to describe three dimensional objects) with oscillating equations (necessary to describe the vibrations of atomic nuclei), a series of equations has resulted which, when combined, describe organic structure. Once the organic structure can be defined mathematically, revelations as to how it can change over time can be postulated. The greatest impact to science and potential breakthroughs will come as three dimensional harmonics are identified which will either stimulate or annihilate complex organic structures within living beings. For example, limb or organ regeneration or the destruction of cancer cells and viruses. To my knowledge, no one has yet proposed this application of the equations.

Now let's throw in another discipline. It's called *Cymatics*. Cymatics owes its origins to physicist Ernst Chladni, born in 1756. Chladni pioneered many of the founding principles in the field of acoustics. He was the first to visually demonstrate that sound waves can produce geometric patterns on a vibrating surface. In 1967, approximately 200 years later, Hans Jenny invented a device known as

89

a tonoscope which used crystal oscillators to control the volume and frequency of vibrations in plates and membranes. Depending on the frequencies generated by his tonoscope, Jenny was able to produce complex patterns in sand particles and water droplets.

So, what does all this discussion mean? As I stated earlier, simple molecules form harmonic chords. As these molecules vibrate they set up structured patterns as demonstrated by Cymatics which, in turn, produce complex molecules. As the molecules grow in size, their frequency patterns change, and they again set up new structured patterns (crystals not required). If you will recall, Dr. Stanley Miller discovered amino acids as well as a number of other molecules necessary for the formation of life in his "Primordial Soup". In chapter 5, I discussed naturally forming molecules which exist throughout our solar system and, in fact, throughout the universe. As the complexity of these molecules increase, they tend to follow a cumulative growth effect known as a golden spiral. (DNA helix anyone?) The golden spiral is formed geometrically using a mathematical series of values known as Fibonacci Numbers.

Know also that the frequency of matter depends on the temperature of its nuclei. The amount of solar radiation that a planet receives from its sun will determine if and when organic structure begins to form. When the harmonics of matter are within a defined set of parameters, life will ensue. The science community has coined the term, "Goldilocks Zone" referring to the not-too-hot, not-too-cold region within a solar system which has the potential to create life.

Specifically, the nuclear resonant frequencies of Carbon, Oxygen, Hydrogen, and Nitrogen coincide at these temperature regions and the Cymatic patterns take effect forming organic structure and, eventually, living creatures. This is intelligent design!

11. All the forces of nature—strong interaction, electromagnetic force, weak force, and gravitational force—are all combined at the "E" state. Separation of the forces occurs as a consequence of space-time expansion voids (vortexes). Solid matter is created when high-energy electrons are wrenched from the compressed proton and neutron within a space-time vortex. This concept is the holy grail of physics. It is the missing piece of the puzzle that scientists have been seeking, and it is completely opposite to prevailing thought.

Simple, isn't it? Perception of the physical universe will either bring further understanding or further frustration. If our perceptions are correct, if we properly apply human facts, and mostly, if we listen to our hearts and to God, we will find the truth. This not only applies to physics but to our relationship with the spirit world as well.

Figure 5

Euclidean Space and
Binary Pairs The Power of 2

(Think Base 8 Mathematics)

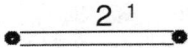

$$2^1$$

Two points in space defining a line or binary pair.
Chemically this would be called a covalent bond.

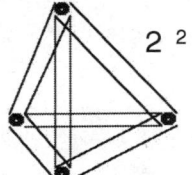 2^2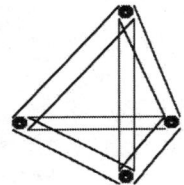

Two tetrahedrons in mirror image or bisection. Together they form half the volume of full space.

Four points in space defining a simple three-dimensional space or a tetrahedron. Many organic molecules are in the form of tetrahedrons.

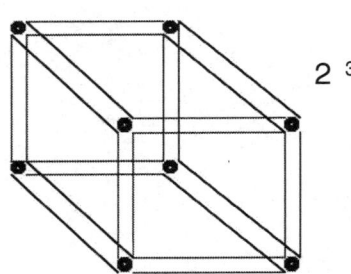 2^3

Note: An octet exhibits four times the volume of a tetrahedron of proportional sides.

Eight points in space defining a cube or full-image space, octet formation.

Chapter Thirteen

And the Answer Is?
History and Education!

Well, I admit, this solution doesn't seem too earth-shattering, but I want to discuss the manner in which I believe these topics should be presented. History, for me, was a pretty boring topic in school. Memorizing names, dates, and places didn't seem like useful information to me since most of these people and places were either dead or demolished, respectively. Occasionally, the teacher would introduce some juicy morsel of information about a famous individual, but I think that was just to keep the students awake. Life as I knew it today didn't seem to have much in common with life of yesteryear. I really didn't care how ancient societies of the world pillaged and plundered their way from one continent to another. Certainly, the people I knew would never behave like this. We were much more advanced. This was a kinder and gentler planet I now lived on. The brutality of our forefathers would never slip its way into modern society. Surely, history such as this would never be repeated!

I think by now you can understand where I am headed with this discussion. As I stated early in this book, life is about cycles. Cycles are repeated. How we deal with the emergence of a cycle that is detrimental to living beings is the key to our future. Recorded history is now of sufficient length that we should be able to recognize when our society is in danger. Unfortunately, just because we record historical events doesn't necessarily mean that we have learned any lessons from their outcomes. Mainstream religions teach their followers lessons of the past by reciting the teachings of David, Mohammed, Buddha, and Jesus Christ. The document that I am most familiar with and that I have used earlier in this text is the Bible. Although it has been translated many times over the ages (sometimes mistakenly so and with omissions), it has provided evidence of our past that

93

we cannot overlook. From a historical perspective alone, the Bible tells stories of greed, avarice, power, decay, lust, war, and death. These stories have significance today simply because these same traits in mankind still exist today. Can we overcome these traits? No. Can we do something about them? Yes.

By profession, I am a commercial airline pilot for a major air carrier in the United States. Since powered aircraft are a relatively new development in the history of mankind, their design and manufacture are still undergoing a great deal of refinement. The entire history of powered flying machines is barely one hundred years. So what have we learned? One of the early things that we learned was that airplanes have a very effective way of terminating human life—they crash. Avoidance of this annoying property became a priority for pilots very early in the program. As time passed and mechanical advancements provided reliability, a new problem began to arise concerning fatalities not associated with equipment failure—human error.

During World War II, when aircraft became a major military weapon, the manufacture of aircraft became one of the world's largest manufacturing undertakings. The nation that could produce the fastest, most reliable, and most maneuverable aircraft would have a distinct advantage over their enemy. Technological advances were a national priority. Since so many aircraft were being built and flown in such a short amount of time, their flaws also became evident at an advanced rate. At the time, new aircraft could be built faster than a pilot could be trained for combat. Safety issues, even in wartime, began creeping into the design phase of aircraft manufacture in an effort to avoid simple mistakes.

Understanding the effects of stress on aircrews and how they handled complex tasks under this stress became a new field of study for aircraft manufacturers. This field of study is known as human factors engineering. One such example of this is in the design of simple controls in the cockpit. Originally, many controls in the cockpits of multiengine aircraft had identical handles and knobs. Although these handles performed vastly different functions, their location in the cockpit was the only obvious difference to the pilot other than a descriptive placard.

Under stressful conditions, pilots were moving the wrong handles to accomplish emergency procedures. The results of these mistakes often resulted in the loss of the aircraft and its crew. The engineering solution was to place critical switches and controls in locations that would not be commonly used, and also to change the shapes of these controls so that the tactile feel would be different to the pilot.

Aircraft design has incorporated the knowledge that humans make mistakes. Aircraft redesign continues to this day to accommodate the mistakes that still occur in our cockpits. Redundancy, auditory warnings, control and knob shapes, location of vital switches, computer databases, air traffic controllers, and additional crew members have all aided in preventing mistakes made in the cockpit. We have learned the frailties of human beings and now engineer around them. Unfortunately, engineers are human also.

The effort continues to perfect the safest aircraft possible. This would not be possible without studying and understanding the history of aviation. As a result, we have been able to avoid the pitfalls of our past. Aircraft accidents continue to occur, but at a significantly reduced rate than the past. The cycle continues, but we have delayed its onset.

So what is to become of humankind without the understanding that we are violent, greedy, and deceitful? How do we engineer our society to avoid these pitfalls? To date, we have accomplished this by building weapons of mass destruction. Just the thought of their widespread use would surely mean the end of all mankind. Many of us are asking, How much longer can this last? When will someone decide that fear is no longer a deterrent to nuclear weapons use? There is an alternative. Knowledge. I would like to provide that knowledge for you. Know the following things:

1. Life is a gift. Treat it as such. Your body is the vessel through which your soul will experience life. Do not abuse your body or let it waste away. Realize that your body came from the earth and will return to the earth many times over. Respect its path through time. You are only the temporary custodian of its matter.

2. Time is eternal, and so is your spirit. This is not your first time on this planet. You don't remember your past lifetimes because you are not supposed to. The trials of life are to be experienced with each dawning of a new generation. Discovery is the wonder of God. It was never meant to be easy.

3. You will be as those who are about you. Love others as you would love yourself. Spirits of God do not know gender or race. Gender and race are physical manifestations of life. All life is equal. This includes the animal kingdom. Do not hold yourself in esteem above others, for you will live as the least of your fellow human beings, or animals, in your lives to come.

4. Respect the nature of the earth. Polluting the water and air is genocide. Nature will retaliate, and mankind will lose. We are as much a part of the earth as it is a part of us. Don't think that we can continue to go unpunished.

5. God's world is cruel to those who will not hear his voice. Look within yourself for truth and wisdom. Pray to those who have left this earth. They will help and guide you. God is made of many spirits who are with you always. You are never alone.

6. Do not worship the physical world. The physical world is only a place of learning, not a place to be acquired. Of course, if you enjoy the accumulation of recombined corpses and manure, then go for it! You can never take it with you, but then, who would want to? Remember, personal wealth is directly proportional to a person's willingness to spend time away from their family, friends, and loved ones. Never trade personal wealth for love and companionship. Find a balance between work and family.

7. Your spirit will return to the earth roughly every two hundred years. Two hundred years equates to roughly ten generations. The effort that you put into your life and the lives of others will be returned to you tenfold. If your efforts are self-centered and self-serving, then there will be little on this world worth returning to. If your efforts are for self-improvement and the improvement of others, then the world you return to will be full of joy, kindness, and fulfillment. What kind of place are you building for your own return?

8. Dedicate your life to the teaching of your children. If we are to be successful as a species, then we must commit ourselves

and our time to the education of our children. An uneducated child is a lost child, and a child without a future loosens the fabric of the woven rug of society.

How easily are the minds of men swayed? When I spoke of education, I did not intend for you to fill your minds with gibberish. I intended for you to fill your minds with the knowledge of science, facts, and engineering. Create something positive for one another, and pass that passion of invention on to those who will follow you in life. Perfecting treachery, deceit, and ill-gotten gain is not a path toward God. Devising plans to reap the rewards of others' hard work is immoral. Governments, organizations, and professions whose primary motivations are to reward themselves from the perspiration and inspiration of others are destructive to the core of self-motivation and self-reliance. Stay focused. Be alert. Society will not advance if those with talent and wisdom cannot be heard or their endeavors nourished.

9. What is the most valuable commodity to living creatures on earth? Time. It is the only thing in our life that can never be replaced, yet it is the one thing in our life that we most take for granted. Only God knows when your work is done. Rejoice in each day that you live.

10. Give your most valuable gift to those about you. Your loved ones require the benefit of your time. Talk to one another. Share your ideas, hopes, and dreams with one another. Take an interest in what your loved ones are doing in their lives. If you have pets, spend time with them as you would another person. Other living creatures can enrich your life in ways that you can never imagine. Quote of the day: Animals do not possess the power to change the world, but they do possess the power to change the hearts of those that do. God's gifts were never

intended to be monopolized by humans. That assumption came from us.

11. The passage of time only occurs at kinetic energies (KE) below the "E" state. Within the "E" state of space-time, there are no todays, yesterdays, or tomorrows. Linear time, as we perceive it, does not exist. Think of galaxies as pinwheels of time, their rotation creating the ticks of time that move our lives forward with each passing day.

12. Scientists have begun referencing thought processes as the "quantum mind", "quantum brain", "quantum cognition", or "quantum consciousness". The latest theories on brain activity and memory concentrate on three dimensional neural functions within the brain; holographic memories, if you will. It has been found that memories are stored within our minds using multiple sources of inputs from our sensory organs such as touch, taste, smell or sound. That is why familiar scents, flavors, or songs trigger memories of other events in our lives. Our minds cross reference memories from previous experiences. But what is brain function? Is it just a bunch of random vibrations within an organic neural network that we can trace on an Electroencephalogram (EEG)? To me it is not.

Scientists of our time prefer to use the words, "quantum states" to separate themselves from dealing with a higher reality. They understand that quantum matter exists in a form of duality (particles and waves), but refuse to admit that heavenly forces are at work in the process.

A) All conscious living creatures store memories as three dimensional images.

B) All memories are engrained at a quantum (read: spiritual) level, referred to as our soul, and are carried with us throughout eternity.

C) We will find that communication can be achieved with other living creatures by mentally projecting images, not words; as words are a human creation. To do so requires a new understanding of ourselves and our potential capabilities.

D) Communication from others in the spiritual realm will manifest as images within our minds. Psychic Mediums have known this for some time, and must interpret the meaning of these images and symbols in order to pass on information in verbal form to a recipient.

E) The driving force of all complex brain activity occurs at the quantum (read: spiritual) level. The soul drives all advanced thought processes and is independent but bound to the physical body. (This concept has escaped scientists who try to connect brain wave activity with the "quantum" effects of neurons)

So now that you know what I know, what do we do about it? The obvious answer is to look introspectively and decide whether you want to make a change in your life and adopt a more holistic approach to living. To do so might involve losing your job and way of life. To not do so might involve condemning your soul to a dying planet and the physical pain and anguish associated with a doomed society.

All I can say is this: God has allowed us the insight to determine our fate. Perhaps this is the special gift that we have received as human beings on planet Earth. Not only can we decide to protect and preserve other species on our planet, but we can decide to protect and preserve ourselves as well. Perhaps in saving others, we will discover the true meaning of the word *saved*. For to be saved is not a choice of the here and now but of the future. A future yet to be lived.

Educating ourselves and our children about God, truth, and love is a good start toward a joyous future. Adopting these lessons into everyday life will give your lives greater meaning and fulfillment. All this translates into peace and happiness. What all this also means is a great deal of effort. The road to knowledge is a rocky one. Never forget that learning requires hours of sacrifice. Hours of sacrifice means time away from other activities that will be, without question, more appealing to you.

Parents have the most demanding job. Leading your children down the path of righteousness and morality will be a matter of your own commitment to God. Children will not follow if they believe that you are not sincere. You must lead your children with an open and loving heart. You will make mistakes along the way as will your children. Always be forgiving to an honest mistake.

Resist the temptation to abandon your education and the education of your children. How can you serve the world without first knowing what the world is about? How can you help yourself or others without developing your talents? How can you serve God and his creatures without an understanding of nature? How will Earth survive if we do not respect its delicate balance?

For the rest of us, we must create an environment that allows parents to do their jobs. Flexible working conditions, reasonable hours of labor, and financial backing of educational institutions will help to ease the financial burden of parents and distribute the responsibility to others in the workforce.

Immoral societies are incapable of self-governance. If you value your freedom, you must first govern your own behavior. Failure to do so will result in a society of brutal laws and oppression designed to break the will of a violent culture. Honor the Ten Commandments. They are the base moral guidelines for Islam, Judaism, and Christianity and contain great wisdom. Interestingly, the *five precepts of Buddhism* contain the same moralities. Abide by them.

The Story of the Great Oak Tree

Once upon a time, there existed a great oak tree. Every fall, the great oak tree would, upon its branches, produce a quantity of acorns so vast as to bend its powerful limbs ever so gently toward the earth. Because of its great bounty of nuts, the squirrels in the surrounding land would flock to its branches and store their winter's harvest into their arbor nests of twigs and leaves. It seemed the great quantity of food would last forever.

The industrious squirrels of the land would toil away for weeks, ensuring that their families would have as much food as they could possibly need throughout even the harshest of winters. Because of this, the lazy squirrels would waste away the cool fall days playing in the branches with their friends and frolicking in the piles of fallen leaves.

As time progressed, the industrious squirrels began to pass away. The lazy squirrels who had not prepared their nests with enough acorns for the winter found themselves going to the nests of the industrious squirrels and asking for additional nuts. This was fine for a while since the industrious squirrels had harvested more than enough for all the squirrels in the tree. But it seemed as the years progressed, there were fewer and fewer nuts to be borrowed. Though the industrious squirrels seemed concerned about this trend, the lazy squirrels were not. They continued to frolic the fall away with nary a care in the world.

It wasn't long before the industrious squirrels began to complain, but it seemed that the ever-increasing numbers of lazy squirrels were not listening. They were content with their lives. Life in the great oak tree was easy and free. Why would it change? Then, one year, winter fell early upon the land of the great oak tree. The industrious squirrels had harvested as much as they could before the great snows. They went to their nests with enough acorns for their own families but that was all.

Late into the winter, the lazy squirrels began to hunger. They went to the nests of the industrious squirrels to borrow food. But since the industrious squirrels were now fewer in number and with less available nuts, there was not much food to go around. Feeling the need to help, the industrious squirrels gave away their remaining stores of acorns to help the lazy squirrels survive. Alas, the winter persisted longer than anyone could have imagined. The food ran out, and all the squirrels of the great oak tree perished.

So what is the moral of the story? Do not live off the fruits of labor of your neighbor. Each is responsible for the harvest of their own acorns in their lifetime. Be content with your own harvest. If you choose to live off the fruits of others long enough, eventually there will be nothing left for anyone.

How do we accomplish such an ambitious undertaking? How do we ensure that the world might be a better place in the future? Priorities!

Chapter Fourteen

The Business of War

In 2011, an estimated $1.7 trillion were spent globally on national defense systems and hardware. Forty-one percent of that spending was done exclusively by the USA; 2.6 percent of the world's gross domestic product (GDP), or roughly $236, was spent for every man, woman, and child on the planet. So how did all this get started?

The Industrial Revolution, which began in Great Britain, spanned the period from 1750 to 1850. It was during this time that great advancements were made in manufacturing and agriculture, specifically the creation of machine tools that were used to produce interchangeable parts.

Around 1850, the Second Industrial Revolution began, which introduced steam-powered ships, locomotives, and power to run large textile mills using refined coal.

By the early 1900s, the internal combustion engine and electric power generation were brought into use. It was not long after that militaries shifted from the use of pack animals to mechanized vehicles.

In August of 1914, World War I began between Germany, Russia, France, and Great Britain. The arms race had officially begun. In November of 1918, the war ended.

When the Nazi party took over Germany in the 1930s, a massive rearmament program began, violating the Versailles Treaty. In September of 1939, Germany invaded Poland, and World

War II began. It was estimated that the total expenditures for the war approached $1.6 trillion. In May of 1945, Germany surrendered.

After World War II, the world powers began to rebuild their manufacturing infrastructures. The United States, which suffered no loss of its manufacturing facilities, began the conversion of its plants to the production of consumer products, such as automobiles and small durable goods. However, companies such as General Motors, General Electric, and Boeing Aircraft Co. continued production and refinement of military hardware, but on a much smaller scale. In 1956, the Federal Aid Highway Act was authorized by Congress to create a national highway system throughout the United States so that military forces could be rapidly mobilized in the event of a foreign invasion.

In October of 1957, the Russians launched Sputnik into low Earth orbit. In 1958, NASA was formed, and the space race between the United States and the Soviet Union began. This was also the same year in which ARPA, or Advanced Research Projects Agency, was created within the Department of Defense. Later, in 1972, the agency would be renamed DARPA. The purpose of which was to identify and counter technological advancements made by our foreign enemies. Science and engineering dominated the quest for military supremacy.

In 1979, the United States and China reestablished normal trade practices. Trade volume between China and the United States went from $2.4 billion in 1979 to $211.6 billion in 2005. This trade volume caused a massive reduction in US consumer-product manufacturing to be transferred to China and other Pacific Rim countries. The United States would be forced to expand the manufacturing of military hardware and increase sales of that hardware to its NATO allies. It was the only avenue left to keep US factories and US jobs viable. Consumer product manufacturing in the United States was, and is, no longer the staple of middle-class America.

When the United States initiated Desert Storm in January 1991, the US economy had benefitted from several years of military expenditures under the Reagan administration and a tax policy that increased revenues to the US government. New and advanced hardware had been developed under DARPA, which was now available for use. This was the first time that the modern world would see how technology had, once again, changed the art of warfare. Laser-guided weapons and beyond-line-of-sight artillery devastated the armies of Saddam Hussein.

In March of 2003, the Gulf War began, and Saddam Hussein was removed from power. Since then, the United States has been involved, militarily, with Iraq and Afghanistan without any perceivable objective. Billions of dollars have been spent and little, in terms of positive outcome, has been accomplished. The US economy has survived due to the aggressive printing of billions of dollars and massive federal spending programs, but debts now exceed $18 trillion.

So was anyone concerned over the growing appetite for military expenditures or the powers they grant to governments? James Madison wrote in 1865, "Of all the enemies to public liberty war is, perhaps, the most dreaded, because it comprises and develops the germ of every other. War is the parent of armies; from these proceed debts and taxes; and armies, and debts, and taxes are the known instruments for bringing the many under the domination of the few."

In January of 1961, Dwight D. Eisenhower warned the American people of the influence of a military industrial complex.

Why am I reviewing our past? Because military expenditures cannot exist unless there is an enemy to fight. James Madison also said, and I quote, "Perhaps it's a universal truth that the loss of liberty at home is to be charged against provisions against danger, real or pretended from abroad." In other words, don't let your government fool you into giving up your freedoms and your hard-earned dollars to fight an invisible enemy.

Political leaders have long used the fear of death and destruction to manipulate their citizens' will. The more a government diminishes the role of a people's faith in God, the more likely they are to deify themselves. Those who put their faith in God rather than a politician cannot be easily swayed to act in an immoral manner against their fellow man.

Since the beginning of the Industrial Revolution, mankind has squandered billions of dollars and hundreds of thousands of lives on tools of destruction for fear of one against another. As I have told you, we all share the universe with God, and our time spent on this earth is not meant to annihilate one another, but to embrace one another.

We face enormous challenges in our near future to ensure the survival of mankind.

Isn't it time that we use what we have learned from the technology of weapons development to save mankind instead of continually trying to destroy it?

Chapter Fifteen

Solutions: What Should the World Be Like in the Future?

Since I have covered so many different subjects throughout this book, I will discuss solutions as individual topics. Are you ready? Here we go.

Transportation. It should be obvious to those of you who live in large metropolitan areas that our current primary form of transportation—namely, the automobile—is creating more problems than it is solving. Rush hour gridlocks last closer to two or three hours. We are incurring a higher incidence of accidents due to a myriad of reasons, some of which are fatigue, frustration, stress, substance abuse, or mechanical problems. Road rage occurs more frequently, resulting in aggressive driving behavior or shootings.

The initial purchase prices of cars are taking a larger percentage of personal income. Insurance rates have followed increased purchase prices, and higher probabilities of accidents occur due to greater numbers of drivers on the roads.

Interstate trucking has increased substantially over the past several years, adding to the congestion and increasing the danger to automobiles and their passengers. Any type of collision between an automobile and semitruck results in severe injury to the occupants of the automobile despite our best efforts to protect occupants with seat belts and air bags. A typical automobile weighs 4, 000 lb. A typical single semitrailer weighs 40,000 to 80,000 lb. $KE = 1/2mv^2$. The energy of the impact is not simply ten to twenty times the weight of the vehicle, but is a function of the velocity squared! Consider a slow-moving automobile up against a fast-moving semi. (You do the math.) Hence, the ensuing carnage.

Fuel prices will not be going down, and consumption continues to go up as vehicular traffic increases. **(In the**

United States, the price for a gallon of gasoline was $1.03 in 1998. In 2011, it was $3.53.)

Our current solutions to the above-stated problems include carpooling, truck routes through cities, HOV lanes, and electronic traffic monitoring and notification signs. All these innovations are fine until some octogenarian decides to pull out onto the interstate at 35 mph in rush hour traffic and initiates a cataclysm of brakes lights, skidding tires, and multicar pileups. The scenarios are endless, but the results are the same. The system grinds to a halt; traffic is backed up for miles, and the delays are intolerable.

Solutions. Regardless of the type of solution we choose, we must abide by the following conditions:

1. The energy consumed to propel our transports must be renewable and must not add carbon molecules to the atmosphere. For example, electrically driven, flywheel driven, or combustion driven (hydrogen fuel cell, for example) vehicles meet this criteria.

2. Mass must be minimized to reduce weight and thus energy consumed. The material must be easy to recycle. For example, plastics and aluminum. Human-powered transport must be utilized to the fullest extent possible. Mass requirements are by far the lowest per individual per mile traveled. Not only that, it's great exercise. Obesity, heart disease, hypertension, and diabetes should greatly decline. General health will improve and, as a result, require lower-calorie diets, reducing the demand for food.

3. The entire system must be expandable and reducible to accommodate shifts in demand and population changes. Railway systems, for example, can either add or subtract passenger cars as the demand dictates. Reducing the number of cars during low-passenger demand requires less energy to be consumed and is therefore more efficient.

4. Safety achievements will occur by utilizing microprocessors to the fullest extent possible. This type of monitoring is cost-effective and energy efficient.

5. The infrastructure must accommodate large pedestrian and human-powered vehicle thoroughfares (sidewalks and bicycle paths) for short-distance connections between mass transit systems.

6. All freight and bulk shipments of raw materials and consumables must be transported so as not to interfere with pedestrian thoroughfares and traffic. The purpose of which is to minimize danger to life and to minimize delays in mass transportation. Minimizing delays in bulk transport also reduces the energy required for such delivery. Separating freight from commuter-travel thoroughfares allows much lighter vehicles to be used for personal transportation. Lightweight vehicles will be just as safe as today's SUVs because there will be no large-mass vehicles sharing the same roadways. Lighter weight means substantially improved energy efficiency.

Energy. It should be obvious by now to even the most casual observer that petroleum, coal, and natural gas will not sustain us for much longer. We can extend the period of their usefulness by conservation efforts, but another solution must be considered. I have suggested that superconductivity is the ultimate solution, but we will need an interim source of energy before this curiosity of physics becomes perfected and put into widespread use.

Solutions. The answer has been staring us in the face for some time now. It began when Stanley Miller defined the ingredients of his primordial soup. Ammonia, hydrogen, and water vapor are the compounds that we must focus on. I have eliminated methane from the list. Why? Because there are no elements of carbon in its composition. The big enemies in our current production of energy are greenhouse gases—namely, carbon monoxide and carbon dioxide. Both are produced by burning carbon-chain molecules. Hydrogen gas is the most promising energy source to satisfy our near and long-term energy needs. Transporting hydrogen is going to be the most challenging transition from fossil fuels. Stabilizing the hydrogen into an inert substance is the key. Diluted ammonia may act as a key transport medium. Although the substance is toxic and slightly corrosive, it is not flammable. Its molecular structure consists of nitrogen (an inert gas) and hydrogen. The two byproducts of ammonia are both desirable candidates as energy sources. Using water to dilute the ammonia provides additional sources of hydrogen and oxygen, which can be separated through electrolysis. Breaking down ammonia into its constituent parts will require a heat source, such as solar energy furnaces.

Housing. Current housing construction in the United States consists of relatively the same materials used for the past 150 years. Because of this, our energy consumption has been rising more each year due to additional electronic devices, central air-conditioning systems, computers, and appliances located within our homes. Prefabricated home sales are booming due to their inherent cost per square foot advantages and less stringent building codes. Although minimum insulation requirements are being met by today's standards, they are far too inadequate for future survival in a world without buried energy reserves.

Solutions. Homes currently known as earthships should be our global goal. These homes consist of recycled materials, such as automobile tires, aluminum cans, glass bottles, active and passive solar radiation panels, batteries, and water storage cisterns. Regardless of the type of materials used in housing, the house of the future must meet the following criteria:

1. All homes must strive to be energy self-sustaining. Only during extremes in climate should homes require supplemental energy for heating or air-conditioning. Transportation will require the most energy in the future. A house, which never moves from its build site, should never require energy to run its systems when it can produce most of what it needs from the sun, earth, or shade. A net-zero house, which has already been demonstrated by Centex Homes in California, would be highly desirable.

Other builders have since joined the trend, but very few have actually been built since I wrote this in 2000. The National Association of Home Builders (NAHB) has, along with others, joined the Department of Energy to develop cost-effective building techniques, which will result in a net-zero home by 2030. (Really? Thirty years to develop an energy-efficient home? We can do better!)

2. All current homes should be retrofit with the most efficient products used in new construction. These new products should include passive hot-water systems, solar panels integrated into roof lines, radiant barriers in attics, geothermal heating and air-conditioning units or water-cooled air-conditioning units, insulated windows made without metal frames (due to their transference properties of heat and cold), reflective paints, programmable thermostats, awnings or modified eves (which maximize shade in the summer but allow sunlight in the winter), sunrooms built with passive thermal mass floors or walls, and low-voltage lighting. Retrofits such as these should have financial incentives to home owners since these costs will be substantial. The alternative, however, would be to live without many of the comforts we now enjoy. By the way, unless we redesign most of the gadgets we cherish so much in our everyday lives, we will only have enough energy in our homes to operate them one at a time.

3. All new home design must be ecstatically pleasing. Integration is the key to all new home construction. Global standards will have to be established. This is not optional. Each nation can choose design parameters and construction materials, but energy consumption is not negotiable.

Since I wrote this text twelve years ago, I never imagined that we would still only be discussing and demonstrating net-zero homes. Recently, I asked a home builder why more people weren't adopting green-energy construction for new homes. His answer was simple. He said, "How much green-energy technology can you afford? Much of what's currently available is not cost-effective and will not see a return on investment for ten to twenty years. People just aren't willing to wait that long."

City planning. Let's face it, we have very little city planning. What we have instead is urban sprawl. The direction of which is determined by the price of land, tax rates, and location of the nearest freeway. Those developers with large investors have the luxury of planning entire neighborhoods and small shopping centers, but that's about it. Although every city has a planning department, most of what gets built is determined by the political clout of the developer and behind-the-door handshakes.

Solutions. Cities of the future will not survive as a wide-open metropolis. Instead they will share the following parameters:

1. Cities will be concentric in design. The centers of the cities will incorporate business centers or manufacturing facilities, shopping centers, grocery stores, and high-density apartment complexes. Streets will fan out from the center like the spokes of a wheel. Residential homes will make up the perimeter of the cities, with occasional mini shopping centers placed for essential living needs. Beyond that, agricultural property will supply cities with perishable goods, such as vegetables and dairy products.

If this design sounds familiar, it should. The blueprint for this design actually began in medieval villages, which existed well before mass transportation. These villages were designed in this fashion out of necessity. This design allows the residents to reach the city center in the shortest time and distance possible. Many ancient cities in Europe still exhibit this design. As a result, Europeans consume much less fuel per capita than their Western counterparts.

2. Walking on sidewalks, riding on bicycles, and riding on personal electric vehicles (such as the Segway) will be the most common forms of transportation in the inner city. Freeways occupy too much valuable space for the residents of concentric cities. Electric vehicles will be allowed, but they will be far smaller than today's automobiles and will utilize much

narrower roads. These roads will be limited to the access of mass transportation centers and food retailers.

Several companies have emerged, particularly in California, that have begun producing electric bicycles (PiCycle) and gyro-stabilized two-wheeled electric automobiles (Lit Motors). Tesla and Fiskars have had products on the road for a few years now. Others are beginning to emerge from Eastern Europe.

Realize that all these changes in our society will provide incredible opportunities for businesses involved in the production and distribution of eco-friendly products. Automobile conglomerates will have an opportunity to produce light rail systems throughout large metropolitan cities. Companies like Owens Corning will be integral in providing insulation products for homes. Phillips, Siemens, Johnson Controls, and General Electric have developed a vast array of electronic control systems and lighting expertise. Apple, IBM, Microsoft, and Sony have the microtechnology software and hardware facilities able to program and run our new infrastructure. We have the manufacturing capability and knowledge to reform our world. The key is pulling it off before we run out of time.

Agriculture. When the supply of fossil fuels is exhausted, one of the first industries to be affected will be the food industry. Farming and ranching have been mechanized for some time now. Most farm equipment is gasoline or diesel powered. With the exception of the Amish, food production will come to a sudden halt without tractors, combines, and pickup trucks. Not only will the production and harvesting cease but also the ability to distribute the crops to the masses. Therefore, the agricultural industries of the world must be the first to adopt alternative-energy conversion.

Solutions. Most existing farms and ranches possess the energy reserves they need to be self-sustaining. Unfortunately, few have taken advantage of these reserves. To date, it has not been economically feasible to make the necessary changes to their operations. There have been exceptions. Innovative dairy farmers and feed lot owners, taking advantage of this energy, have already demonstrated that they can produce enough methane gas to power all their machinery and heat their homes by harnessing the natural decomposition of plant and animal waste. These readily available resources come from unused plant material and animal manure. Collecting this material, placing it into large storage containers with expandable bladders, and introducing bacteria into the mix, creates large volumes of methane gas. The methane gas is then routed to converted diesel generators which then supply electricity to the entire operation. These innovators hold the key to our future survival. Nations blessed with fertile lands will be the food and energy suppliers to our future world. Countries that promote energy self-reliant farms and ranches will most easily weather the storm created by exhausted fossil fuels.

In addition to farming operations, several cities have begun harnessing methane gas from their landfills which, in turn, are used for either incineration purposes or power production. I would also recommend combining sewage treatment plants with power plants; the by-products of the first providing raw materials for the second. (Human waste to energy)

Defense—national and personal. As I mentioned in chapter 13, mankind has delayed the onset of global war by developing weapons of mass destruction—nuclear bombs. The use of such would cause irreparable damage to our atmosphere and annihilate most life forms on our planet.

Solutions. These weapons will continue to exist in the arsenals of advanced nations until such time we as human beings recognize our true place on earth. God will not intervene if we choose to exterminate ourselves. This is the lesson we must learn. Ridding the world of nuclear weapons will not eradicate armed conflict until all of mankind recognizes the common eternal spirit within each of us. Overcoming individual religious beliefs and adopting a holistic view of the world is the only long-term solution. Until that occurs, each nation will have to support a defensive military force. Enlightenment will not occur overnight. Personal defense is another matter. We in the United States have taken on the same philosophy of personal defense that we have on national defense. Extreme-force weapons preclude conflict. If we arm ourselves with the most destructive guns and bullets, then no one can harm us. Therefore, everyone must possess lethal weapons to defend themselves. We justify this logic by quoting our constitutional right to bear arms. Unfortunately, the framers of the Constitution gave the public the right to bear arms in order to defend themselves from a tyrannical government. Two hundred years ago, the federal government had the same weapons to defend the nation that each citizen had to defend themselves from one another. How long do you think it would take the federal government to defeat a public uprising today? Each member of the federal armed forces takes an oath that, in part, states, "I will defend the Constitution of the United States from all enemies foreign and domestic." Any uprising would be squelched in a matter of hours. Like it or not, the right for the public to bear arms in today's world serves a significantly different purpose. That purpose is only for the defense of oneself and one's family. I agree that we all have a right to defend ourselves with personal weapons. So why do they have to be lethal weapons?

Solutions. Law enforcement agencies across the nation have been utilizing nonlethal weapons for riot control for many years now. These weapons and many more under development must be the only weapons available for purchase by the general public. Why? Because it will save lives. These should be our priorities for the future:

1. Gun manufacturers must develop nonlethal defensive weapons for use by the general public. An example of this, as yet to be invented weapon, would be a capacitor discharge/neurotoxin gun; a weapon combining the effects of a Taser and a tranquilizer gun. Stored on a nightstand in a power charger, it could be fired at short range using a CO_2 cartridge. Selectable settings on the pistol grip would give the user the option of either jolting or incapacitating a target, depending on the threat. A

light emitting diode on the tip of the barrel, either red or green, would signal the individual being targeted of your intensions. Coincidental with the development of these types of weapons would be the cessation of sales of lethal firearms to the general public. This includes the ammunition to fire them.

2. As more defensive weapons find their way into the marketplace, the federal government will purchase all privately held firearms. All such firearms must be destroyed in a timely manner.

3. The eradication of rifles must also be part of the plan. Hunters will need to hone their skills with tranquilizer guns and compound bows if they still insist on killing members of our animal kingdom. I think it would be much more sporting if the hunters were at equal risk of losing their lives to that of the animals. Kill or be killed. Now that's hunting!

4. As more of us come to realize that life is a gift from God, we will not feel the need to defend ourselves with lethal weapons. Who among us would want to explain how a loved one was killed by mistake because we were in fear of our life? Who wouldn't be afraid knowing that millions of lethal firearms were in the hands of paranoid citizens in a heavily armed society? Fear will never be a deterrent. The lack of fear is the true solution.

I know that this type of thinking is a bit altruistic and maybe a bit naive, but I still believe that mankind, given the chance, can meet the challenge. As I like to say, I am a pacifist; I will always be a pacifist; and I will defend that position to the death.

Overpopulation. Our current population increase is driven by many factors. Firstly, it is driven by the mechanization of agricultural equipment coupled with the fossil fuels used to power them. Secondly, with the development of nuclear weapons, war has become a thing to be avoided. Thirdly, medical advancements have caused the life spans of millions of people to be extended. Without war, famine, and disease to reduce population growth, we face inevitable decisions. These decisions not only include the human population but the animal population as well.

As the human population encroaches into undeveloped areas of the world, indigenous species are left to fend for themselves. In most developed countries, predators, which are a danger to humans as well as other species, have been eradicated. This leaves animals, lower on the food chain, free rein to reproduce at will. Many communities are faced with large populations of deer, elk, squirrels, rabbits, mice, and rats, to name a few. Our current solution is to declare open seasons on many of the larger species in order to control their proliferation.

Solutions. Controlling birth rates to coincide with mortality rates is the only long-term solution. Achieving this objective should be accomplished using similar techniques for both the human and animal populations. Controlling birth rates is a matter of controlling the mating pairs.

Science fiction writers have been delving into this subject for decades. Their solutions include the euthanasia of people at age sixty (*Soylent Green*), eating people to feed an underground race (*The Time Machine*), applying to the government for permission to breed (*Brave New World*), or fighting simulated wars and vaporizing the virtual casualties (*Star Trek*). Which one would you choose? The governments of all countries will have to decide the best course for its citizens. The more populated the county, the more drastic the measures will need to become. Don't be surprised if this suppression of procreation causes a mass exodus from heavily populated countries to less populated countries. Just remember, the sooner we get started reducing world population, the less horrific the transition will be.

Controlling the populations of large wild animals such as deer and elk will be much more straightforward. Tranquilizing the females, inserting a Norplant device, and reintroducing them to the wild will be a good start. Castration of the stags is another but more costly and dangerous solution. Continuing to allow the use of lethal weapons to indiscriminately murder these animals is brutally unnecessary. Unless and until it becomes necessary to kill these animals to ensure our own survival, there is no need to casually destroy their lives.

Deciding the total number of beings allowed to inhabit our Earth will be the most difficult decision of our lifetimes. Satellites, orbiting Earth, have been mapping and collecting data for some time now. We have the information at our disposal today to determine how much animal and plant life our world can accommodate. Estimating survivable biomass without the use of fossil fuels will be the challenge. Avoiding mass starvation in a world without buried reserves will be a monumental feat.

The cure for cancer. The cure for cancer of all kinds will occur with the adoption of radio-frequency therapy, or RF therapy. What is RF therapy? Well, it doesn't exist yet. What does exist is a machine called an MRI, or nuclear magnetic resonance imagery. This machine creates a strong magnetic field that encases the patient on a motorized bed. Using a combination of strong magnetic fields and tuned radio frequencies (usually corresponding to hydrogen atoms), the resulting effect on human tissue is the creation of radio-frequency waves. These radio-frequency waves are captured and displayed on a monitor. The resulting image is a color representation of the internal organs, muscles, bones, and tissues of the patient. For cancer patients, these images also reveal the location or locations of tumors. Each structure within the human body emits a distinct radio-frequency pattern. This frequency pattern is in reality several frequencies on many wavelengths. These wavelengths represent the molecular structure of the object being imaged. Remember earlier I stated that molecules form a vibrational chord. RF emissions from an MRI are reflections of those vibrational chords. Here is the importance to cancer therapy. If it can be imaged, it can be isolated and destroyed. Disruption of the cell mitosis of cancer cells can be accomplished using RF dissonance. Technology to create RF dissonant transmitters will follow the same path used in the development of noise-cancellation headsets used by airplane pilots. The theory is the same. Location-specific transmitters, strength, and frequencies of the RF transmissions will be the most difficult challenges. The more frequencies that can be matched to a tumor's cellular makeup, the more effective the therapy and the quicker the recovery of the patient. The specific tuning of frequencies will be critical in eliminating unwanted tumors, viruses, and bacteria within our bodies. Chemical compounds currently used to combat many diseases and ailments will no longer be necessary.

Hold the presses! Since writing these paragraphs, it has come to my attention that RF therapy was developed in the 1930s by *Dr. Royal Rife* as a treatment therapy for cancer patients. The Rife frequency generator was designed to identify the "mortal frequency" of many tumors and viruses. It seems that Dr. Rife claimed significant success using this technology. These generators are still manufactured today and are endorsed by individuals in naturopathic medicine. Mainstream medicine has not endorsed this technology, and for obvious reasons (*money* and pharmaceuticals), they are not likely to any time in the near future. What I find most fascinating is that I reached the same conclusions as Dr. Rife concerning the reaction of cancer cells to radio frequencies without a knowledge of his research. The major difference in my proposal for fighting disease is the utilization of multiple frequency harmonics, which would increase the effectiveness of the treatment dramatically. Current Rife frequency generators do not. As a result, the success rate for curing ailments is relatively low.

Education. I've spoken a great deal about the importance of education in modern society. All great nations of the current world share a commonality. They all have wonderful institutions of higher learning. As it turns out, the most advanced nations are not those with the most natural resources but rather those with the highest percentage of educated citizens. It must be a national mind-set and a national priority stressed to every generation in order to succeed. Consider countries like Japan, Germany, Denmark, Sweden, Norway, Canada, or Australia. There is a distinct difference between the "haves" and "have nots" in our world today. That difference is education. Any subculture within a society that does not adopt education as a necessity of life, both mentally and spiritually, will find itself financially and morally bankrupt. I am not a strong proponent of technology training in early development at schools. Music, art, and literature must be reintroduced as an alternative to computer labs and computer programming classes. Creativity and expression of emotion must go hand in hand with learning the fundamentals of math and science. Artistic expression is what keeps us in tune with our senses and our soul. It is also something that we can share among ourselves from generation to generation. Think of all the great artists and musicians throughout the centuries. Michelangelo, da Vinci, Monet, van Gogh, Dalí, Picasso, Mozart, Beethoven, Schubert, Liszt, Led Zeppelin (just kidding; well maybe), etc. It is these kinds of endeavors that keep us focused and grounded. In order for the world to survive as a whole, we must adopt a global strategy for education.

1. We must create a world education fund. The purpose of which is to construct standardized schools and classrooms in third world nations. Help those to help themselves. Along with these schools must come dedicated teachers and governments willing to back them. Educated citizens will find solutions to their own country's problems, create businesses, provide jobs, and eliminate the need for intervention by advanced nations. Every individual as well as every nation must be self-sustaining. Pouring cash and food into starving nations only prolongs an already-desperate situation and does not provide a long-term solution. The cash usually ends up in the hands of those in power, and the food ends up being confiscated and sold on the black market.

2. We must break the chain of illiteracy in subcultures of advanced nations. Highly developed industrial nations will collapse without a strong base of educated citizens. We can no longer hide behind our racial and ethnic differences. Find a way to love those different from yourself and build new relationships that will eliminate the rifts between us. All citizens of a nation are responsible for its success or failure. Parents must get involved with their child's education. Too much emphasis has been put upon the educational institutions to guide and motivate young students. This shift of responsibility does not work! Stop placing

blame for failure onto everyone but yourselves. United we stand, divided we fall.

3. We must incorporate a sense of stewardship for our planet into every new generation of citizens. As global communication expands into the everyday lives of more individuals, we find ourselves in a seemingly much smaller and accessible world. Humanity is becoming much more aware of its effect on the rest of its inhabitants. The time required to share that knowledge is dwindling day by day. We can get the world to change for the betterment of all life. We now have the tools to make it happen. Every human capable of reading and writing must be a part. Don't be afraid. God is on our side.

4. Finally, we must find the resolve within ourselves to accomplish all these tasks. Take it upon yourself to make the changes in your life that are beneficial to yourself and to others. Each person on earth must earn their right to exist and grow. With individual effort should come individual reward. Fostering a competitive nature among nations is a healthy endeavor. Mankind has made the greatest strides when forced to do so. National pride is essential, so long as is does not result in international conflict. The first nation that develops complete energy independence based on solar derived sources wins. They will reap the financial rewards by selling their technology to the rest of the world.

Use what you need of this earth, and leave behind more than you started with. You will be rewarded in spirit by those you have helped in soul.

Chapter Sixteen

And in Conclusion

While serving in the USAF, I was asked by a fellow air crew member if I thought traveling into space was important to humanity. My response was, "Yes, I think that it is vitally important. How else will we find out that all the mysteries of the universe could have been solved here on Earth? We will find that we never had to leave." What did I mean by this? I meant that the most important things in our lives are here before us. The time that we share with the ones we love, including our friends, family, spouse, children, and pets, is irreplaceable. This moment in time, the experiences we share, the places that we live will never occur twice in all eternity. This particular mix of matter and DNA will not occur again.

The energy of the spirits of planet Earth will remain with planet Earth. Should one traverse the known galaxy and lose one's life on some remote foreboding planet, one's spirit will be trapped among the life energies of that planet forever.

There is evidence that a society that once inhabited Earth was very much aware of this phenomenon. Not until recently have we come to realize that the great pyramids of the Egyptian Giza plateau were related to the heavens. The configuration and size of the pyramids have a direct correlation to the stars located in the belt of the constellation Orion.

The shape of the pyramids has been a puzzlement. Consider this possibility. The shape of the pyramids is not unlike the shape of a glass prismatic structure. Glass prismatic structures have the property of dividing pure white light (wide-range electromagnetic waves) into their constituent color bands. An electromagnetic wave diffuser, if you will. The diffusion occurs because the light energy is directed from the outside, not the inside. Originating the energy source from the interior of the great pyramids will have the opposite effect—phasing, alignment, or concentration of the energy source.

Located within the centers of the pyramids are burial chambers. From the burial chambers to the exterior surface of the pyramids is a narrow passage. The passage is aligned with the ancient celestial location of the stars of Orion's belt. I believe that the purpose of the pyramids was to capture and phase the spiritual energy of the "traveler" entombed within and direct this energy back to its rightful place. This was twelve thousand years ago, roughly the time when the fabled continent of Atlantis perished. Since then, several cultures have copied the shape of the pyramid, with little understanding of its purpose. These mimicked structures usually exhibit many staircases on their sloped sides and have altars atop their apexes. Obviously, mankind could not grasp the purpose or understand the function of a spiritual wave phaser. I'm not so sure that we will today. Perhaps a visit to the planets in Orion's belt is in order. (Update: NASA's Kepler space telescope has been tasked with finding potentially habitable planets located within the Milky Way; specifically in the stars making up Orion's Belt!)

The unique experience that is human being is specific to Earth, just as it is for other beings on other planets. In other words, if you choose to explore the galaxy, be very, very careful. Don't trust a society of intelligent beings unless they share the same universal truth as yourself. What is the universal truth? Haven't you been paying attention? All life is equal. All life comes from God. God is us, them, and everything in between. God is the Alpha and the Omega.

Spirits within God have been trying to make contact with those of us in the physical realm for countless generations. Events that defy explanation have been described by thousands of individuals, only to be criticized or ridiculed by the disbelievers. In current times, people seem to be more open to spiritual encounters having actually occurred. Reality television shows have attempted to bring validity to the phenomena. Spiritual mediums, such as John Edward, Char Margolis, and Theresa Caputo, have all been featured on various television networks by sharing their gifts with the masses.

What I want to share with you is this: Mankind has reached a point in its evolution to accept that there is an invisible universe waiting to be discovered. It is the next big step. Technically speaking, paranormal events teem around areas of low electromagnetic disturbances. From the evidence gathered, most of these events occur in the dead of night, when the sun has the least effect on the ionosphere. The manipulation of the electromagnetic spectrum is the easiest to manipulate by spirits of God. In addition to electromagnetic waves, spirits also influence electromechanical devices, such as switches and motors.

Our thought processes can also be influenced by the spiritual realm. Dream states, in particular, can be used by spirits to visit loved ones. The purpose is to give a sense of peace to the relatives still living after having experienced a devastating loss.

Notice that there is a common thread to all these phenomena. They all occur at the transitional energy state from solid matter to energized matter. The ripples of space-time are the communication frequencies of the here and now to the hereafter. It is the elusive connection to God that we have been unable to comprehend, until now.

Several years ago, the now defunct magazine *Omni* had a contest among its readers that challenged them to write a short letter entitled "What Would You Say to an Alien?" I decided that day to write down my thoughts. I never submitted my work to the magazine but instead kept it to myself. I never read the winning letter. I suppose I wanted to think that mine would be published in another venue someday. I guess I didn't think anyone would understand where my version came from anyway. Now, I'll let you decide.

WHAT WOULD YOU SAY TO AN ALIEN?

Friend,

Welcome to Earth. If you can read and understand this message, then with this welcome must also come a warning. The bipedal society inhabiting this planet has a history of violent and destructive behavior. Beware. Few on this world believe or respect the existence of living beings beyond this planet. Protect your knowledge. It will be ridiculed. Protect your technology. It may be duplicated and used against you. Those of us within our society who are prepared for your arrival are very pleased. We have been waiting a very long time. Although you are free to remain here as long as you wish, do not protract your stay. It will take many years for our society to assimilate your existence into our philosophy of life. Hopefully, over time, we will cease our behaviors of aggression and oppression and, instead, replace them with tolerance and compassion. When you return, perhaps we can share time together in our solar-sustained homes, eat fresh vegetables from our private gardens, sip herbal teas, and stroll along pristine lakes, observing the natural beauty of a spring day while feeding wild deer from the palms of our hands. Perhaps together, we can look into the heavens and profess our mutual love, joy, and awe of life itself.

Your friend,
Craig

Thoughts after
September 11, 2001

Many people over the years have asked me to what religious affiliation I belong. I tell them that I have no affiliation at all. Most people react by asking me if I am an atheist. I tell them that I am not an atheist but instead consider myself deeply spiritual. They then inquire as to where my spirituality came from and how can I just make up whatever I choose to believe. I tell them that my beliefs are guided by God. They ask me, "Where is that written?" or "Where did you read that?" They take offense to someone choosing their own path to God rather than following an organized religion.

For all of you who consider yourselves devout followers of an organized religious group, I have this to say: once that you have chosen a religion, you have chosen to take a side. You have categorized yourself into a way of thinking that makes inevitable comparisons with other religious doctrines and rituals of life. You demean and ridicule cultures that have adopted other religious belief systems. You attempt to change or convert these poor souls into thinking as you do. You distrust them. You attempt indoctrination.

All this is an unfortunate consequence of being social beings. Other social species on our planet suffer the same affliction. They congregate with others similar to themselves. They shun others with dissimilar hair color, body shape, or behaviors. These traits are outside of their comfort zone. There is an unwillingness to coexist, so they exist independently of one another. They are on opposing sides. What we all witnessed on September 11 was precipitated by the fear and mistrust that exist between religious cultures on our planet. The incursion of religious beliefs or cultural behaviors into unwilling nations was met with ultimate and unyielding force. It was a reaction of fear. It was a reaction of desperation. It was a reaction of mistrust. It was a reaction of ignorance. It was an attempt to snuff out those different in religious doctrine and cultural behavior.

Imagine what it would be like for a small nation with limited resources, steeped in ancient beliefs, to be confronted with an overpowering wave of foreign wealth, technology, and diametrically opposing philosophies of life and priorities. Imagine that your entire culture is in jeopardy of losing its identity and history. Imagine the influence of wealth and power on a people with none. Imagine the possibility of losing your relationship with God to the power and influence of materialism and hedonism. Imagine that these influences rock the very foundation of everything that you have believed in for your entire life. Imagine that this influence is, in reality, great evil. Imagine facing the Great Satan himself. Would you trust the Great Satan? Would you bargain in good faith with the Great Satan? Or would you fight the Great Satan to the death, believing that you were protecting yourself and your loved ones from the overwhelming power of all that is unholy and evil? Imagine that you believed your fight to be endorsed by God himself and you would be rewarded in heaven by defeat of the Great Satan.

If you can imagine all these things, then you can comprehend the power and destruction behind defending religious differences. You can comprehend the dangers that we face as a world, divided among ourselves, in the name of God. Who are we? We are us and they are them.

The time could be a thousand years ago or a thousand years from now. It was September 11, 2001. When will we learn?

Author's note: The Great Satan to which I refer is the name that Muslim fanatics have given to the United States of America. Their holy war, jihad, is a perceived fight between good and evil. Only men and women can choose to do good and evil acts. God cannot. Therefore, the campaign is immoral and contrary to God's will. Self-defense against this aggression seems justified, but without establishing a common ground for peace, the conflict will plague our world for years to come. We must seek common wisdom in our mutual belief in one God. How we glorify his name in our disparate faiths must not turn us one against the other.

As you can see, not much has changed since I wrote this chapter in 2001.

Post Log:
Personal Thoughts From
the Author

Mankind's greatest achievement has been and will be debated for many decades. Inventions such as the steam engine, the telephone, the lightbulb, or the television have been cited as prime examples. I personally believe mankind's greatest achievement to date is global communication. We are still struggling to find the best utilization of this communication network, and time will reveal if it truly is our greatest achievement. It is my hope that through the use of global communication, I can send a message of hope and understanding unto the world in which peace will emerge.

In case you haven't made the connection, this philosophy (truth) I am offering to the world is the solution to world order and peace. It is the merging of science and religion. Both, over the millennia, have sought to reveal the truth of our existence. So far, both have failed. Mankind has chosen, for whatever reason, to live outside the boundaries of nature. We have separated ourselves from the living, breathing organism that we call Earth. We have found this existence to be independent of all other living creatures. We are wrong to do so.

Mankind has developed different religious doctrines so ingrained into our cultures that we are still willing to murder one another over our differences. We are wrong to do so.

Mankind has developed weapons of such magnitude that deciding to use them against one another would not only eradicate the human species but all others as well. We do not have the right to decide for the nonhuman species of this planet. We share our planet. We do not own nor shall ever own this world.

Mankind, to this day, does not know what its purpose is or should be as stewards of Earth. I must tell you now that we have

131

lost our way. Can we find a solution? Yes, but you (society) will have to grow up. I am speaking of social maturity. You will have to begin to accept that all life comes from God and all life returns to God. You will have to open your heart to the possibility that what I have tried to explain to you in this book is the truth. The truth is that life is the coexistence of both the physical universe and the spiritual universe.

Why haven't we come to this conclusion already? Firstly, because religious doctrines stagnated hundreds of years ago and have failed to accept scientific change and, secondly, because scientists choose to research the physical universe without regard to its spiritual nature. Had both groups done so, world philosophy would have reached the same conclusions I have.

There will be no world peace without first disarming yourselves of the emotional weapons of bias and hatred toward others of different faiths, cultures, races, or gender.

There will be no world peace without world education. Ignorance and fear go hand in hand.

There will be no world peace without care and compassion for all living creatures. Accepting other creatures as equals on this earth will ensure their survival and our humility.

There will be no world peace without universal law and order. The Ten Commandments are a good start.

Human beings will always have to be kept under constant scrutiny. We know that we are basically untrustworthy. Why this is so, I have no idea other than to suggest that perhaps we consider this an essential survival trait.

Should this book be widely received, I expect to receive severe criticism from scientists and religious organizations across the globe. I expect that my life will be over as I currently enjoy it. I do not desire fame and fortune. Neither buys too much within the spirit of God. I do desire change for the world.

We have reached a technological point in our evolution to make great strides in science and social reform. We can do so with the sharing of global knowledge. We can share that global knowledge with our network of global communication. This is how we shall achieve global education. This is how we shall achieve global peace. This is how we shall enjoy the miracle of life for generations to come for all nations of the world.

It seems I was not far off the mark when I stated the importance of global communication. Unfortunately, as with all technology, mankind must make a choice in how best to use it. When mankind first discovered fire, who would have thought that we would be able to harness its properties to heat our homes, power our vehicles, or propel us to outer space? On the other hand, we can just as easily strike a match, cover our world in oil, and burn the whole thing to ashes. It's really up to us. Playing video games with other people around the world, online pornography, or texting your friends about your latest escapade is not quite the application I was hoping for ten years ago when utilizing global communication hardware. Tools are only as good as the people who use them.

Take my advice. Put down your computer, take a walk outside, and let your senses absorb as much of Mother Nature as you can stand. When you get back, hug your loved ones, tell them you love them, and thank God for your life.

Glossary of Terms, Names, and Phrases

<u>Names</u>

Aristotle, 384 BC-322 BC. Greek philosopher who studied at Plato's Academy and later founded his school, the Lyceum. The school emphasized biology and natural philosophy. Aristotle was noted for developing a hierarchy of species, the predecessor of evolution, 2,200 years before Darwin. He also suggested the fifth natural element, aether, which made up the heavens.

Bernoulli, Daniel, 1700-1782. Swiss mathematician and medical doctor who came from a family of noted mathematicians. He developed a fluid dynamics law that related pressure, density, and velocity. The consequence of this law is that if the velocity of a fluid increases, then the pressure of the fluid falls. The law also applies to gases and is the principle for which lift is produced on an airfoil. In 1737, he published *Hydrodynamica*.

Copernicus, Nicolaus, 1473-1543. Polish astronomer whose education was primarily in medicine and Greek. He also held a degree in canon law. In 1513, he wrote a short work known as the Copernican theory. This theory placed the sun at rest in the center of the universe and is referred to as heliostatic cosmology. In 1543, just before his death, *On the Revolutions of the Heavenly Spheres* was published.

Darwin, Charles, 1809-1882. British naturalist who traveled the world aboard the HMS *Beagle* from 1831 to 1836. In 1859, he published his most notable book titled *The Origin of Species by Means of Natural Selection, or the Preservation of Favored Races in the Struggle for Life.*

Einstein, Albert, PhD, 1879-1955. German physicist who, in 1905, wrote *On the Electrodynamics of Moving Bodies*. Within this paper was what became the special theory of relativity. He received the Nobel Prize in Physics in 1922. In 1939, he and other scientists proposed to President Franklin D. Roosevelt the possibility of making an atomic bomb. It was considered likely that the German government was already pursuing this endeavor.

Galileo Galilei, 1564-1642. Italian mathematician and astronomer. He developed the telescope in 1610 and discovered and catalogued the mountains of the moon, the moons of Jupiter, the phases of Venus, and other heavenly bodies. He heavily favored the Copernican world view and came into conflict with the church when he published *Dialogue Concerning the Two Chief Systems of the World—Ptolemaic and Copernican*. In 1632, he was condemned to indefinite imprisonment but eventually was confined to his villa until his death in 1642.

Jenny, Hans, 1904-1972. Swiss physician and natural scientist. Noted for his work in the study of wave phenomena, Jenny published two volumes of work on Cymatics. The first volume, *Cymatics: The Study of Wave Phenomena,* was published in 1967, and the second volume was published in 1972. Jenny, considered the father of cymatics, was heavily influenced by the work of Ernst Chladni who, in 1787, discovered the effects of sonic frequencies on the formation of patterns in sand. Jenny expanded this work by including the cymatic effects on fluids and pastes demonstrated on a device, which he invented, known as a *tonoscope*.

Kepler, Johannes, 1571-1630. German mathematician who, with the work of Tycho Brahe, published *Astronomia Nova* in 1609, which established the first two laws of planetary motion. He was the first to suggest that the planet Mars exhibited an elliptical orbit around the sun.
In 1618, he discovered the third law of planetary motion. He is also credited with many other firsts including the title of founder

of modern optics, the principles of the telescope, magnification, internal reflection, and inverted images. In mathematics, he formed the basis for integral calculus and mathematically derived logarithms.

Maslow, Abraham H., PhD, 1908-1970. American psychologist who developed a hierarchical theory of human needs composed of deficit needs (or D-needs) and a hierarchy of information and being needs (or B-needs). The D-needs were as follows: (1) biological/physiological, (2) security/safety, (3) social, (4) ego/esteem, (5) self-actualization/fulfillment.

Meissner, Gerald W., PhD, 1941?-present. American physicist. He is currently professor of physics at the University of North Carolina at Greensboro. He is responsible for demonstrating magnetic levitation above a superconducting field commonly referred to as the Meissner effect.

Michelson, Albert A., 1852-1931. Polish-born American physicist who, in 1878, was the first to accurately measure the speed of light. He and his colleague, Dr. Edward W. Morley, used interferometry in an attempt to discover the existence of an ether (or aether). It is commonly referred to as the Michelson-Morley experiment. Because the experiment failed to prove that an ether existed, Einstein's special theory of relativity, which offered an alternative explanation for the propagation of light, was considered correct.

Miller, Stanley L., PhD, 1930-present. American biochemist who, in 1953, designed a laboratory experiment that resulted in the formation of amino acids without free oxygen. The amino acids glycine, alanine, aspartic, and glutamic acid, among others, were produced.

Newton, Sir Isaac, 1643-1727. English mathematician and physicist. He is most noted for his work in physics and celestial mechanics. His early work focused on the theory of light and color. He was the first to suggest that white light was composed of several different light rays. In 1687, he published *Philosophiae Naturalis Principia Mathematica*, or *Principia* as it is most commonly known. From this book led him to conclude the law of universal gravitation, which states, "All matter attracts all other matter with a force proportional to the product of their masses and inversely proportional to the square of the distance between them."

Planck, Max Karl Ernst Ludwig, PhD, 1858-1947. German physicist who, in 1900, proposed a formula now known as Planck's radiation formula. This theory proposed that electromagnetic radiation traveled in quanta. The quantum theory of energy was later verified by Neils Bohr in 1913 by using the formula to calculate positions of spectral lines. Planck received the Nobel Prize in Physics in 1918.

Rife, Dr. Royal R., 1888-1971. American scientist and inventor who, in 1931, developed an electromagnetic therapy device designed to kill microorganisms. In 1934, he claimed to have cured sixteen cancer patients using this technology. His invention was dismissed by the American Medical Association as quackery and was suppressed by various private and government agencies.

Terms and Phrases

five precepts of **Buddhism**: (1) Avoid taking the life of beings. (2) Avoid taking things not given. (3) Avoid sensual misconduct. (4) Refrain from false speech. (5) Abstain from substances that cause intoxication and heedlessness.

aether (also ether). (1a) The rarefied element formerly believed to fill the upper regions of space, (1b) the upper region of space, heavens; (2a) a medium that in the undulatory theory of light permeates all space and transmits transverse waves, (2b) the medium that transmits radio waves.

big bang. A theory in astronomy: the universe originated billions of years ago from the explosion of a single mass of material so that the pieces are still flying apart.

condensation nuclei. Small particles in the air created from/ by dust, volcanoes, factory smoke, forest fires, ocean salt, and sulfate particles from phytoplankton in the oceans. These particles are required to form cloud drops in the troposphere.

cymatics: In physics, the study of wave phenomena, especially sound waves, and their effects on various forms of matter. These sound wave patterns have also been tested on human subjects, for the purpose of holistic healing.

dark energy. A hypothetical form of energy that permeates space and exerts a negative pressure, which would have gravitational effects to account for the differences between the theoretical and observational results of gravitational effects on visible matter. Dark energy is not directly observed but rather inferred from observations of gravitational interactions between astronomical objects, along with dark matter.

dark matter. A hypothesized form of matter particle that does not emit or reflect electromagnetic radiation (planets, asteroids, cosmic dust). The existence of dark matter is inferred from gravitational effects on visible matter such as stars and galaxies.

exobiology. The search for life on other planets currently being funded through NASA.

fetal alcohol syndrome (FAS). Infant medical disorders associated with a mother's consumption of alcohol during pregnancy. FAS is characterized by (1) abnormal facial features, (2) growth deficiencies, and (3) central nervous system problems. FAS is a permanent genetic condition.

Möbius loop. A one-sided surface that is constructed of a long rectangle by holding one end fixed, rotating the opposite end 180 degrees, and applying it to the first end.

morphic fields. A term coined by British biochemist Rupert Sheldrake, PhD. As the author of over fifty papers published in scientific journals, Dr. Sheldrake has pursued the concept of spiritual guidance and awareness in living creatures. Many of his books suggest experiments that attempt to verify paranormal experiences such as ESP.

Noah's ark. According to the Bible, Noah constructed a large boat, known as an ark, to house his family and pairs of animals. The ark saved the group from flood waters caused by forty days and forty nights of rains. Once the Great Flood receded, as described in the book of Genesis, the ark came to rest on the mountains of Ararat.

prebiotic. An early phase in the emergence of life on a planet. It is during this phase that volcanic activity, lightning discharges, and solar radiation begin the formation of key molecules such as sugars, amino acids, and nucleotides. These molecules are the building blocks of proteins and nucleic acids.

primordial soup. Also known as the soup of life. It is the prebiotic mix of simple molecules proposed by Dr. Stanley Miller.

Over the past fifty years since he began his experiments, Dr. Miller and his team at UCSD have continued to modify their mixture to obtain additional organic molecules essential to the development of life.

quintessence. The fifth and highest element in ancient and medieval philosophy that permeates all nature and is the substance composing the heavenly bodies, the essence of a thing and most concentrated form.

singularity. A point where some property is infinite. To extrapolate the properties of the universe to the instant of the big bang, you find that both the density and the temperature go to infinity.

40112522R00094